Modality: Issues in the Semantics-Pragmatics Interface

Current Research in the Semantics/Pragmatics Interface

Series Editors:
K.M. Jaszczolt, University of Cambridge, UK
K. Turner, University of Brighton, UK

Related Elsevier books

In this Series:

TURNER (ed.)	The Semantics/Pragmatics Interface from Different Points of View
JASZCZOLT	Discourse, Beliefs and Intentions
GEURTS	Presuppositions and Pronouns
JASZCZOLT (ed.)	The Pragmatics of Propositional Attitude Reports
PEETERS (ed.)	The Lexicon-Encyclopedia Interface

Related Elsevier journals

Journal of Pragmatics
Editor: Jacob Mey

Language and Communication
Editors: Roy Harris and Talbot J. Taylor

Language Sciences
Editor: Nigel Love

Lingua
Editors: Johan Rooryck and Neil Smith

Free specimen copies available on request

For further information on the CRiSPI series and for details of how to submit a proposal go to:
www.elsevier.nl/locate/series/crispi

MODALITY: ISSUES IN THE SEMANTICS-PRAGMATICS INTERFACE

Anna Papafragou
University of Pennsylvania, USA

2000
ELSEVIER

Amsterdam - Lausanne - New York - Oxford - Shannon - Singapore - Tokyo

ELSEVIER SCIENCE Ltd
The Boulevard, Langford Lane
Kidlington, Oxford OX5 1GB, UK

First edition 2000

Library of Congress Cataloging in Publication Data
A catalog record from the Library of Congress has been applied for.

British Library Cataloguing in Publication Data
A catalogue record from the British Library has been applied for.

ISBN: 0 08 043634 X

∞The paper used in this publication meets the requirements of ANSI/NISO Z39.48-1992 (Permanence of Paper).
Printed in The Netherlands.

Current Research in the Semantics/Pragmatics Interface (CRiSPI)

The aim of this series is to focus upon the relationship between semantic and pragmatic theories for a variety of natural language constructions. The boundary between semantics and pragmatics can be drawn in many various ways, the relative benefits of each gave rise to a vivid theoretical dispute in the literature in the last two decades. As a side-effect, this variety has given rise to a certain amount of confusion and lack of purpose in the extant publications on the topic.

This series provides a forum where the confusion within existing literature can be removed and the issues raised by different positions can be discussed with a renewed sense of purpose. The editors intend the contributions to this series to take further strides towards clarity and cautious consensus.

Contents

ACKNOWLEDGEMENTS

This book grew out of my doctoral dissertation (Papafragou, 1998a) which was completed at University College London. I owe a great debt to my main supervisor, Deirdre Wilson, for her wisdom, enthusiasm and support, her sharp judgement and the enormous generosity with which she has shared her time and thoughts with me. Her enduring insights have influenced my thinking in innumerable ways, and her views on communication and cognition have provided the theoretical background on which the present work is based.

Neil Smith, my second supervisor, has been an unfailing source of insightful comments and provocative questions. His comments were always delivered with incredible wit and invariably uncovered a deep knowledge of linguistics.

The Department of Linguistics at UCL is an exciting place to study pragmatics and communication; this book owes a lot to work within relevance theory carried out by members of the faculty and fellow graduate students. My own research has profited greatly from discussions with and comments from Robyn Carston, whose specific interests have consistently provided me with inspiration and motivation. Villy Rouchota has offered me illuminating opinions and invaluable help on a number of occasions. Other members of the Department contributed in various ways to the shape and content of this book: Annabel Cormack, Dick Hudson and Rita Manzini offered different perspectives and many insightful discussions.

Much of the material in this book was originally presented to the UCL Pragmatics Reading Group. I want to thank Marjolein Groefsema, Steve Nicolle, Vlad Zegarac, and my fellow research students Richard Breheny, Corinne Iten, Eun-Ju Noh, Milena Nuti and George Powell for helpful feedback. I also thank Keith Brown and especially Ruth Kempson for their comments and encouragement.

In 1998-9, I visited the University of California at Berkeley as a postdoctoral researcher. During this time I had the opportunity to present and discuss various aspects of this work and receive many suggestions from a number of people in and outside Berkeley. Special thanks go to my host Eve Sweetser, and to Kent Bach, Alison Gopnik, and Dan Slobin for comments and discussion. I also received helpful feedback from Ayhan Aksu-Koç, Eve Clark, Jill de Villiers, Nina Hyams, Stephen Neale, and Elizabeth Traugott.

The last stages of writing the manuscript took place at the University of Pennsylvania during the first year of a postdoctoral fellowship at the Institute for Research in Cognitive Science. I benefited greatly from the lively interdisciplinary atmosphere of the Institute, and I am especially grateful to Lila Gleitman for inspiration and support from the early days of my arrival. I also wish to thank Robin Clark, Henry Gleitman, Aravind Joshi, Tony Kroch, Alan Leslie, Ellen Prince, Maribel Romero, and the members of the CHEESE seminar for energetic conversations on semantics, pragmatics, language development and everything in between.

I would like to offer personal thanks to Georgios Babiniotis, Dimitra Theofanopoulou-Kontou, Keti Bakakou-Orfanou, Dimitra Chila-Markopoulou, Amalia Mozer, Christoforos Charalambakis, Kiki Nikiforidou, Stella Vosniadou, Irini Warburton-Filippaki and Ad Neeleman for their guidance and moral support. They have taught, challenged, inspired and encouraged me at various stages of my academic life; without them, maybe I wouldn't have become a linguist and this book wouldn't exist.

Large portions of this material have been presented in front of audiences from linguistics and psychology in various conferences and departments, including the University of London, University of California at Berkeley, University of Pennsylvania, University of Lancaster, University of Cardiff, University of Hertfordshire, UNAM (Mexico City), Max Planck Institute for Psycholinguistics, University of Athens, University of Thessaloniki, Emory University, and Boston University. I am grateful to the participants of these talks for many critical points and suggestions, a good number of which have found their way into the final form of the book. Earlier versions of this material appeared as articles or book chapters; I have included the relevant references in appropriate places in the chapters to follow.

For providing financial support, I would like to thank the State Scholarships Foundation in Greece, the A. G. Leventis Foundation, and the Institute for Research in Cognitive Science at the University of Pennsylvania.

I have been fortunate to have had lots of friends in Britain, Greece and the States who made life great fun in the past few years; I thank them all. George Pappas has been a wonderful source of inspiration and support; he will always have a special place for me in all possible worlds. Most of all, I thank my parents, Dimitris and Eleni Papafragou, and my sister Efie, for their loving support, trust and understanding. They were always there for me, even though being far away. This book is affectionately dedicated to them.

1

INTRODUCTION

1.1 THE ISSUES

1.1.1 Context and Word Meaning

A well-known property of lexical items in natural language is that they are capable of conveying different meanings in different situations of utterance. Examples of the context-dependence of lexically expressed meanings include the following:

(1) a. The lawyers approached the *bar* to have a word with the judge.
b. The lawyers approached the *bar* and ordered two martinis.

(2) a. I asked her many times but got no *answer*.
b. I rang her many times but there was no *answer*.

(3) a. I want to fly like a *bird*.
b. A *bird* was flying above the corpses.

On a traditional analysis, examples (1)-(3) correspond to distinct semantic options (see Cruse, 1986; Lyons, 1977; Saeed, 1997). In (1), *bar* is lexically ambiguous between (roughly) the senses 'court area' and 'area serving drinks': it therefore corresponds to two distinct (and unrelated) entries in the mental lexicon.[1] In (2), *answer* is polysemous: it encodes two separate but related meanings, and is thus treated as a case distinct from (1).

[1] Notational conventions: Concepts are cited in capital letters. The use of small-case italics is generally reserved for lexical items. I will occasionally use single quotation marks to refer to an expression's meaning.

In (3), *bird* is semantically univocal but its contextual interpretation takes on different properties depending on pragmatic considerations. (3a) involves the construal of a member of the category BIRD which is very close to the prototype: something like a swallow; (3b), by contrast, involves an exemplar which is closer to a vulture. One of the main problems for linguistic theory has been to account for the various aspects of the context-dependence of lexically communicated meaning - which includes (but is not limited to) the phenomena in (1)-(3) - in a principled way.

Of the three possibilities above, polysemy has always been the hardest to delineate, and its territory the most difficult to separate from either ambiguity or semantic univocality (monosemy). Interestingly, the concept has been revived and developed in much recent research within philosophy, cognitive psychology, linguistics and artificial intelligence, where there has been a resurgence of interest in the structure of the mental lexicon and the way it interacts with other components of cognitive architecture (see selectively Nunberg, 1979; Jackendoff, 1983, 1990; Lakoff, 1987; Levin and Pinker, 1991; Nunberg and Zaenen, 1992; Sag and Szabolski, 1992; Gibbs, 1994; Goldberg, 1995; Pustejovsky, 1993, 1995; Tsohatzidis, 1990; Lehrer and Kittay, 1992; Pustejovsky and Boguraev, 1996). Although there is still little consensus among different researchers as to how the subvarieties of systematic multiplicity of lexically conveyed meaning are to be captured, many of authors converge on the conclusion that our lexical entries have a rich internal structure, which often over-determines the output of the comprehension process (in the sense that it furnishes more than one candidate sense, between which pragmatic interpretation has to choose). This structure is polysemic,[2] in that it consists of clusters of related concepts which correspond to different contextual readings of a lexical item. Details differ among accounts, and polysemy is often used as a portmanteau term to cover a variety of cases - with corresponding implications for the division of labour between this phenomenon and either ambiguity or monosemy.[3]

---

[2] I will use the adjectives *polysemous* for words/lexical items, and *polysemic* for the clusters of interrelated meanings themselves (cf. also *monosemous* vs. *monosemic*).

[3] For instance, polysemy has been invoked to describe and explain: the selection of different complements by aspectual verbs such as (i); the effects of the direct object on the meaning conveyed by the verb in (ii); the effects of the modified noun on the adjectival modifier in (iii); 'semi-conventionalised' uses such as (iv); 'predictable sense extensions' such as (v):

(i) I *began* the novel/reading the novel/to read the novel.

(ii) They *opened* the car/the bottle/the envelope.

(iii) My boss has a *healthy* face/diet/dog.

(iv) Artistic creation is a *miracle*.

(v) *The third violin* is playing badly.

On *begin* and *healthy*, see Pustejovsky (1993, 1995), Pustejovsky and Boguraev (1993), Pustejovsky and Bouillon (1996); on *open*, see Taylor (1995); on 'semi-conventionalised' uses and 'rule-governed sense

This study explores aspects of the context-dependence of lexically communicated meaning focusing on the specific case of modality in English. Given the flexibility of lexical concepts, one would want to know what it is about their semantic representation that allows them to receive multiple interpretations in context - or, to put it differently, what sorts of starting points are provided by grammatical information and how they are exploited by the pragmatic mechanisms of utterance comprehension. The chapters to follow argue for the role of pragmatic factors (and against polysemy) in explaining the plasticity of lexically conveyed information in modal expressions. The next section briefly presents the issues raised by the context-sensitivity of modal items and motivates the choice of modality as a testing ground for settling border disputes between lexical semantics and pragmatics.

1.1.2 The Problem of Modality

Modal expressions allow us to talk (and modal concepts allow us to think) about states of affairs which are not present in the current situation and may never occur in the actual world. In the linguistics literature, it is widely acknowledged that modal expressions may be used to communicate at least two broad clusters of meanings: *epistemic* modal meanings, which roughly deal with the possibility or necessity of an inference drawn from available evidence, and *deontic* modal meanings, concerned with the necessity or possibility of acts performed by morally responsible agents, e.g. obligation and permission (Lyons, 1977; Kratzer, 1981a; Palmer, 1986, 1990). The utterances in (4) and (5), on their preferred interpretations, are examples of epistemic and deontic modality respectively:

(4) a. You must be John's wife.
b. It may rain later in the afternoon.
c. Bioethics lectures should prove interesting.
d. Might I ask whether you are using the typewriter?
e. The sea ought to be visible from the balcony.
f. It is possible that the universe keeps expanding.

(5) a. Employees must feed the animals twice a day.
b. Whoever has finished may go.

---

extensions', see Apresjan (1973), Ostler and Atkins (1992), Copestake and Briscoe (1996), Fauconnier (1985, 1997). I return to examples of this type in section 1.2.1 below; see also 5.2.

c. You should be grateful to your parents for their support.
d. This book might serve as a good introduction to Picasso.
e. He ought to do as she says.
f. It is not possible to get a refund for used beddings.

Apart from the epistemic/deontic distinction, another main area of modal meaning is often recognised, *dynamic* modality, which includes the notional categories of real-world ability, possibility and intention/willingness (von Wright, 1951; Palmer, 1990):

(6) a. Ned can speak four languages.
b. When I was younger, I could run two miles every morning.
c. It is possible for multinational corporations to keep production costs very low.
d. I will become the best skier in the world.

A fourth class of modal interpretations includes *alethic* modality; this has been the traditional concern of logicians and philosophers and deals with absolute or logical necessity or possibility (von Wright, 1951):

(7) a. This conclusion is not necessarily false; in other models, it is actually true.
b. It must be the case that two plus two equals four.

Alethic modality belongs to the conceptual family of epistemic notions and can therefore be distinguished from deontic and dynamic modality. As for the last two types, they are normally grouped together under *agent-oriented* modalities (to be distinguished from *speaker-oriented*, i.e. epistemic, modalities - cf. Bybee and Pagliuca, 1985; Bybee *et al.*, 1994; Bybee and Fleischman, 1995a), or *root* modalities (Hofmann, 1966; Bybee, 1988a; Sweetser, 1988, 1990; Traugott, 1989). In the chapters to follow, I will generally adopt the root-epistemic distinction throughout; I will occasionally refer to dynamic modal readings as *simple root* interpretations.

As the above examples show, it is commonly the case in English that a single modal expression is capable of conveying both root and epistemic modal meanings. As Joos (1964: 195) put it, 'within the modal system English does not distinguish between duty and logic'. Interestingly, the same phenomenon seems to have a robust cross-linguistic presence: sample examples from other languages are given below (see also Fleischman, 1982; Perkins, 1983; Traugott, 1988; Traugott and König, 1991; Bybee *et al.*, 1994):

(8) German

a. Er muss bleiben.
He must-3pres.sing. stay-inf.
'He must stay'.

b. Er muss geheiratet sein.
He must-3pres.sing. married be-inf.
'He must be married'.

(9) Modern Greek

a. Prepi na figis.
Must-3pres.sing. to go-2pres.sing.subj.
'You must go'.

b. Prepi na ine megaliteros mu.
Must-3pres.sing. to be-3pres.sing.subj. older than me
'He must be older than me'.

(10) Gaelic (Adger, 1997)

a. Feumaidh thu seo a dheanamh.
Must you this A do-VN
'You must do this'.

b. Feumaidh gu bheil thu trang.
Must that be-pres.dep. you busy
'You must be busy'.

(11) Tamil (Palmer, 1986)

a. Kantacaami vantaalum vara-laam.
Kandaswami come-concess. come-perm.
'Kandaswami may perhaps come'.

b. veeŋum-ŋŋaakkaa naalekki avan peeca-laam.
Want-cond. tomorrow he speak-perm.
'If he wants, he can speak tomorrow'.

(12) Colloquial Cairene Arabic (Palmer, 1986)

a. Laazim tiXallas bukra.
Must you-sing.finish tomorrow
'You must finish tomorrow'.

b. Laazim jikuun hinaak.
Must he be there
'He must be there'.

Naturally, it is not the case that the forms used for root and epistemic modality are always identical in every language. For instance, in English, it is doubtful whether *can*, compared to *may*, has any genuine epistemic uses. An indication of the restrictions in the interpretation of can is given by the pair in (13):[4]

(13) a. I may have a pencil.
b. *I can have a pencil.

Moreover, there are various constraints on the distribution of modal interpretations, including notorious idiosyncrasy in the interaction of modality with negation (see chapter 3). Nevertheless, the root-epistemic alternation seems to be characterised by a sufficient amount of cross-linguistic regularity to deserve an explanation. Given the pervasiveness of the root-epistemic shift, the question naturally arises: What is the source of this alternation? Is it due to semantic encoding or to pragmatic aspects of utterance interpretation? Are modals lexically ambiguous, polysemous or unitary in meaning? More generally, what underlies the variability in modal interpretation including the subtle sub-types within the root and epistemic classes? Do natural languages grammatically distinguish among dynamic and deontic, or epistemic and alethic categories of modal meaning? Accordingly, how does semantically specified content interact with pragmatic mechanisms to yield the various types of modal interpretation in context? Consider the examples:

---

[4] Also: In Modern Greek, the verb *boro,* when denoting epistemic possibility occurs only in the 3rd person singular as an 'impersonal' verb (with no subject agreement); however, it appears with fully inflected forms (and subject agreement) in its root senses:

(i) a. I fitites bori na figun avrio.
The students *boro*-3sing.pres to leave tomorrow
'The students may leave tomorrow'.
b. I fitites borun na figun avrio.
The students *boro*-3pl.pres. to leave tomorrow
'The students can leave tomorrow'.

In colloquial Cairene Arabic, different forms are used for epistemic and root possibility (Palmer, 1986):

(ii) a. Ti?dari tifuuti min hina.
You-sing.can/may you.sing.pass from here
'You can/may pass through here'.
b. Jimkin jikuun hinaak.
Probable/probably he be there
'He may be there'.

(14) a. As a former champion, John can lift heavy weights.
b. As a simple guest, John can dress casually.
c. As a University employee, John can get health benefits.
d. As a human being, John can have conscious mental states.

Intuitively, these examples convey different types of possibility (including physical, social, legal, biological, etc.) depending on a network of appropriate background assumptions. In explaining how the semantically encoded meaning of *can* and context conspire to produce the range of readings in (14), an adequate theory of modality should address the nature of these background assumptions and explain when and how such assumptions are recovered and monitored during verbal communication.

1.1.3 Scope and Objectives

My goal in this book is to contribute towards the characterisation of the context-dependence of modal expressions and, in doing so, further illuminate the nature of the interaction between lexical semantic information and pragmatic inference. I focus mostly on a representative sample of modal verbs in English (mainly *must, may, can, should,* and secondarily *ought to*). My basic claim is that the semantically encoded content of these verbs grossly underdetermines the interpretations which the verbs receive during utterance comprehension. More specifically, I argue that the English modals have unitary semantic content which, in conjunction with different pragmatic considerations, gives rise to an array of distinct contextual readings. If this characterisation of the semantics of modals is correct, it follows that previous attempts to describe the process of modality interpretation in English in terms of purely structure-driven algorithms are misguided. It also follows that whatever aspects of context have a bearing on the determination of modal interpretation should be handled by the same principles which govern context selection for utterance interpretation in general.

A major component of the proposed analysis is what I call the *metarepresentation hypothesis* for epistemic modality. According to this hypothesis, epistemic expressions are markers of a logical relation between the complement of the modal and the set of the speaker's beliefs. Consequently, the appropriate use and comprehension of epistemics crucially presupposes the ability to conceive of evidential relations between propositions which form the content of beliefs, and is thus linked to the human capacity to metarepresent. The main original contribution of the study lies with the contention that the metarepresentation hypothesis, together with minimalistic semantic assumptions and independently motivated pragmatic considerations, offers a way of handling a broad and diverse range of previously unrelated and puzzling facts and is therefore preferable over previous accounts of modality. Two main classes of phenomena are used to illustrate and motivate this thesis.

In the first place, I consider a variety of arguments which have standardly been taken to provide difficulties for unitary semantic analyses of English modals – and have therefore been seen as grist for the mill of ambiguity or polysemy analyses. These arguments involve the syntactic properties of different modal interpretations, the behaviour of modal readings with respect to truth-conditional content, and the existence of modal constructions with fixed and non-transparent semantic and pragmatic properties. It turns out that a unitary semantic account of modality coupled with the metarepresentation hypothesis can account for these facts in a simple and elegant way. More specifically, I show that the alleged grammatical reflexes of the root-epistemic distinction should be treated as natural aspects of the interpretation process rather than semantically specified features determining the distribution of root or epistemic readings. Moreover, I demonstrate that what appears to be non-truth-conditional behaviour in epistemic modality falls out from the metarepresentational properties of epistemic complements plus independent pragmatic considerations. Furthermore, I claim that the category of speech-act modality (accepted by some researchers as a separate, 'constructional' type of modality) can be reanalysed as a distinct sub-type of the metarepresentational use of language. Consequently, none of these domains can offer reasons to abandon the univocal semantic analysis of the English modals.

In the second place, I consider how the proposed account meshes with data from the acquisition of modal interpretations by the language-learning child. I provide evidence for the claim that epistemic modality involves mentalising abilities and show that its development correlates with other aspects of human metacognitive abilities. This approach correctly accounts for the priority of root over epistemic interpretations in language acquisition and traces some intriguing connections between early linguistic and conceptual development. Data from a cognitive disorder (autism) are also considered in the light of the metarepresentation hypothesis about epistemic modality. Finally, I briefly address the historical development of the root-epistemic alternation and argue that the link between epistemic modality and metarepresentation allows fresh insights into the diachronic facts.

The ultimate objective of this study is to separate the contribution made by linguistically encoded information and inferential processes in the derivation of contextually attested interpretations of lexical items. These two sources of linguistically communicated information essentially correspond to the semantics-pragmatics distinction, at least according to one way of drawing it (see section 1.2.1). In the last part of the study, I examine the implications of the analysis of modality for competing approaches to the organisation of the lexicon. In doing so, I give special consideration to the concept of polysemy as it has been employed in recent linguistic theory. My contention is that this concept is not always homogeneous enough to qualify as a distinct structural type of lexical meaning; moreover, its use within linguistic theory often presupposes some questionable assumptions about the structure of human concepts. I go on to propose a division of labour between ambiguity, semantic underdeterminacy, and a narrowed version of polysemy, and present its ramifications for the psychology of word meaning.

There is by now a huge literature documenting the context-dependence of modal expressions in a variety of languages and attempting to deal with some aspects of the phenomenon in semantic, syntactic or pragmatic terms; several of these works will be discussed in the main body of the book. The intricacy in the behaviour of individual modal items on the one hand, and the very diverse theorising invested in modal data on the other, can make the enterprise of constructing a coherent and explanatory analysis of modality seem formidably elusive. There are two points I wish to make with respect to the empirical scope of this study. Firstly, the present approach does not set out to provide a fully comprehensive account of the state of affairs in the English modal system. Consequently, my analysis is quite compatible with there being a certain degree of idiosyncrasy in modal interpretations, so that some uses of modals have to be listed in the English lexicon and memorised individually. The main argument in this book remains unchallenged to the extent that the main descriptive and explanatory project in dealing with the semantics of modality in English can be carried out without *primarily* appealing to separately encoded clusters of meaning. Secondly, the English case study represents only one of the possibilities which languages possess for drawing the line between encoded and inferred information in modal expressions. As I showed in the previous section, different linguistic systems pose different constraints on the distribution of the root-epistemic alternation. However, it is my contention that this cross-linguistic variation can be explained in part by appealing to the broader semantic and pragmatic framework proposed here for modality in English. The picture which emerges from the following pages is thus intended as a conceptual map of modality, within which it is possible to locate cross-linguistic similarities and differences.

Any linguistic investigation into modal categories is bound to verge on traditional psychological and philosophical concerns with the role of modal concepts in thought and reasoning. From the standpoint of psychology, theories of the lexicalisation and use of such concepts in natural language raise important issues regarding the relationship between linguistic knowledge, inferential capacity and broader cognitive systems in humans. I will address these issues (albeit indirectly) in connection with the treatment of epistemicity and metarepresentation. My approach stands in greater distance from the philosophical enterprise of developing distinct systems of modal logic.[5] I assume that the types of inference warranted by different modal concepts (and investigated by deontic, epistemic etc. logics) do not immediately translate into arguments for the semantics of modal expressions in natural language. This position contrasts with several analyses of modality within formal semantics and philosophical logic which take for granted that variations in modal interpretation should be captured in terms of distinct logical forms (Cross, 1986; Kamp, 1978; Groenendijk and Stokhof, 1975). For such analyses, a treatment of modal expressions in terms of ambiguity seems to be a solution both plausible and natural. From the perspective of the present study, modal expressions in natural language mostly express incomplete or otherwise underspecified contents; these contents need to be pragmatically manipulated to yield the rich modal concepts which formally support inference. In this sense, my treatment is closer to the accounts of Lewis (1986), Wertheimer (1972) and Kratzer (1981a, 1991), which view modal expressions in natural language as underspecified quantificational devices.

1.2 Theoretical Commitments

In developing a semantic and pragmatic account of modality, I adopt the relevance-theoretic approach to utterance comprehension. In the following pages, I give a brief summary of basic relevance-theoretic tenets; for fuller accounts, I refer the reader to Sperber and Wilson (1986/1995, 1987), Blakemore (1992).

1.2.1 The Semantic Underdeterminacy Thesis

According to relevance theory (Sperber and Wilson, 1986/1995; Wilson and Sperber, 1988b; Carston, 1988, 1998; Blakemore, 1987, 1992), (linguistic) semantics corresponds (broadly speaking) to knowledge of meaning which is provided by the grammar; more particularly, lexical semantics characterises the linguistically encoded content of lexical items. This view of linguistic semantics is resolutely cognitive. It presupposes a

---

[5] On modal logic, see Hughes and Cresswell (1968).

representational model of mind (basically of the Fodorean type; see Fodor, 1975, 1983, 1990), which is based on the following premises: (a) the human cognitive system perceives and processes information from the environment through the construction of propositional representations in the language of thought; (b) language is an input system, a specialised module in the architecture of the mind, which is geared towards the translation of natural-language strings into conceptual representations. Grammar, according to this picture, is a mapping between natural-language forms and conceptual representations which constitute the logical form of an utterance. Logical form is thus the output of a specialised and autonomous module and can be plausibly seen as a formula in a mental language.

On the level of individual lexical items, the linguistic mapping onto logical form essentially corresponds to the 'opening up' of the conceptual addresses of individual constituents of an utterance. According to Sperber and Wilson (1986/1995: 86ff.), the decoding of single lexical items typically gives access to a conceptual address, which provides further access to a tripartite informational structure.[6] This structure consists of the following kinds of sub-entries: (i) a logical entry, which includes a set of deductive rules that apply to the logical forms of which the concept is a constituent; (ii) an encyclopedic entry, which contains information about the objects (events, properties) which instantiate the concept; (iii) a lexical entry, which contains information about the natural-language item which encodes the concept. Logical entries can be represented in terms of Fodor's meaning postulates, i.e. elimination rules. (15) gives the elimination rule for the concept GIRAFFE and (16) the elimination rule for the concept YELLOW (examples from Sperber and Wilson, 1986/1995: 92):

(15) Input: (X - GIRAFFE - Y)
Output: (X - ANIMAL OF A CERTAIN SPECIES - Y)

(16) Input: (X - YELLOW - Y)
Output: (X - COLOUR OF A CERTAIN HUE - Y)

[6] I concentrate here on those kinds of linguistic decoding process which contribute a concept to the logical form of the utterance, since this work focuses exclusively on conceptual semantics. Relevance theory recognises a second type of linguistically encoded content which contributes a procedure, rather than a concept, to the interpretation process; procedural semantics thus constrains some aspect of the inferential search for the intended interpretation of an utterance. On procedural semantics, see Blakemore (1987, 1988), Wilson and Sperber (1993a), Rouchota (1994a, b), Rouchota and Jucker (1998), Breheny (1999).

Encyclopedic entries capture most of our knowledge about categories. For instance, our encyclopedic entry for LONDON contains information about the city's geographical location, history, population, and present character; this information is organised in terms of frames, schemas, or scripts, and may contain the representation of prototypes, i.e. typical exemplars of a category (e.g. long-necked animals with brown patches which live in Africa for the category GIRAFFE above - for details see Barsalou, 1992). Lexical (or linguistic) entries contain information about the syntactic, phonological and other properties of the natural language expression which encodes the concept. Occasionally, some of these sub-entries may be empty or lacking in particular concepts: for instance, the concept encoded by *and* has no extension and hence no encyclopedic entry.

On this model, then, the output of the grammar-driven process of decoding a linguistic string (i.e. the logical form of the utterance) will be a configuration of conceptual addresses. A crucial claim made within relevance theory is that this logical form grossly underdetermines what the speaker has said (*the semantic underdeterminacy thesis*). In order to arrive at the *proposition expressed* by the speaker's utterance, the hearer has to perform a series of operations which lie beyond his semantic, or indeed linguistic, competence.[7] Evidence for this underdeterminacy position is supplied by the following example:

(17) Mary said she will leave this place in the evening.

(17) illustrates a host of phenomena which involve the intrusion of non-linguistic parameters into utterance interpretation. The logical form of the utterance underdetermines the full proposition which the speaker intended to communicate in a number of distinct ways. To sample a few: the hearer has to assign reference to the proper name *Mary*; he has to interpret the indexical expression *this place* in a way which will crucially draw on the current situation of utterance; he has to decide on the temporal co-ordinates of *in the evening*; finally, he has to choose between two scope possibilities for the prepositional phrase *in the evening*, which result in different interpretations of (17). Reference assignment and the resolution of indexicality, spatiotemporal coordinates and scope belong to the most widely recognised cases of semantic underdeterminacy: they clearly show that the logical form of the utterance is typically schematic in many respects - what Sperber and Wilson (1986/1995) call an *assumption schema*, which serves as a starting point for the construction of what was said by the speaker.

---

[7] Throughout the book I will generally refer to the speaker as female and the hearer as male.

The development of the underdetermined logical form of the utterance into the proposition which the speaker intended to express is effected via a process of pragmatic inference. Pragmatics thus starts where linguistics leaves off, and that can be quite early in the verbal comprehension process. In order to flesh out the incomplete logical form of the utterance, the hearer unavoidably needs to compute the speaker's intention in producing the utterance, and this in turn imposes specific constraints on both the steps and the outcome of verbal understanding (I will examine these constraints in some detail in the next section). Generally, the hearer takes the logical form as a basis for constructing a hypothesised representation of the proposition which the speaker intended to express (the *propositional form* of the utterance); he then proceeds to the confirmation/rejection of his interpretive hypothesis. The whole process is a case of non-demonstrative inference: i.e. there is no fail-safe algorithm for the computation of the propositional form of the utterance out of the initial assumption schema delivered by linguistic knowledge.

The endpoint of the inferential development of the logical form of the utterance is a complete logico-conceptual representation. It is this form which will receive a truth-theoretic interpretation, i.e. it will be put into correspondence with states of affairs in the world (or possible worlds). The placement of the elements of a mentally represented formula in correspondence with an 'external' reality lies at the core of most truth-theoretic approaches to semantics: on the picture I have sketched here, it may be seen as a distinct enterprise (which can be called 'the semantics of mental representations') which inevitably complements the translational workings of the grammar.

From the point of view of a cognitive account of utterance comprehension, then, the distinction between semantics and pragmatics essentially coincides with the distinction between linguistic decoding and inference. On the relevance-theoretic view, pragmatics uses as a basis the information provided by the grammar to arrive at the endpoint of the interpretation process (retrieving in the process a variety of communicated assumptions other than 'what the speaker said', e.g. intended implications; see 1.2.2). This way of drawing the semantics-pragmatics distinction is by no means uncontroversial (see Grice 1975, 1989; Bach 1994a, b; Gazdar, 1979; Levinson, 1983, 1999; Turner, 1999; for a review, see Carston, 1998). What does seem clear, however, is the extent to which grammatically delivered information needs to be supplemented by non-linguistic input in order to arrive at the truth-evaluable propositional form of the utterance which the speaker intended to communicate.[8]

---

[8] Some version or other of the semantic underdeterminacy thesis is widely accepted in the literature: see Fauconnier (1985), Travis (1981, 1985), Recanati (1989, 1993), Searle (1983, 1992), Atlas (1989), Pinkal (1995), Partee (1995b). Carston (1998) contains a wealth of references and detailed discussion from a relevance-theoretic point of view.

The fact that natural language utterances vastly underdetermine the thoughts they are used to communicate (even on the basic level of 'what was said') is not accidental or trivial; indeed, it is quite plausible in evolutionary terms that, given a powerful inferential system, reasons of cognitive economy would motivate the use of information encoded by natural language expressions as a starting point for the retrieval of the concept/conceptual representation the speaker had in mind. In non-linguistic intentional communication this capacity is exploited quite regularly: a persistent stare or rolling one's eyes may convey very different assumptions in different situations.

The claim that what is communicated goes well beyond what is linguistically encoded sets a research agenda for particular cases. Given the idiosyncrasy of lexical information and the diverse evolutionary origins of our concepts, there is every reason to suppose that our lexical concepts are far from uniform; accordingly, the procedures whereby the information made accessible by a single conceptual address may be moulded into a certain shape in on-line communication may be far from uniform too.

With this background, we may separate different varieties of the flexibility of linguistically expressed (lexical) concepts. Consider the utterances:

(18) a. They *cut* the cake/the lawn/the paper/the bread.
b. My friend has a *fast* car/typist/mind.
c. This steak is *raw*.
d. Juliet is *the sun*.

(18a) offers an instance of a univocal term which can have different interpretations depending on its direct object. *Cut* may be seen as semantically indeterminate: it exhibits a degree of vagueness/sense generality and may require contextual enrichment (or narrowing down) in order to convey a more specific concept. This type of pragmatic interference with the concept encoded by *cut* may be called *free enrichment*, since it is not transparently triggered by linguistic factors.

(18b) illustrates a second kind of semantic indeterminacy, which may be more accurately described as *semantic incompleteness*: expressions such as *fast* include in their semantic representation a sort of gap or slot, which needs to be filled in contextually. Scalar adjectives such as *fast, rich*, indexicals (*we*, *here*, *now*), demonstratives (*that*, *this*) or items such as *local, recent*, etc., require *saturation*[9] (rather than free enrichment) of their informationally incomplete semantic entry: in (18b), depending on the type of modified object, the scale of fastness is correspondingly adjusted.

[9] For the term 'saturation', see Recanati (1993).

(18c) does not involve semantic indeterminacy of any sort, since the semantically encoded content of *raw* is narrow and definable (*raw* means 'not cooked'). The context-dependence of the example comes from the different ways of pragmatically broadening (or *loosening*) the encoded content of *raw* so as to convey a rough approximation (cf. also *silent room, square jaw, straight hair*). According to relevance theory (Sperber and Wilson, 1985/6, 1986/1995; Carston, 1996), when a concept is used loosely, (part of) its logical entry is dropped, whereas the concept retains some of its most salient encyclopedic properties: a silent room is not perfectly soundproof but still allows concentration, isolation, etc.

Finally, (18d) is a more creative case of concept loosening: the metaphorical use of *sun* is at the end of a continuum which includes ordinary cases of broadening the encoded content of a concept so as to cover instances not previously present in the denotation of the concept. This example, then, essentially works like the previous one; the difference lies in the richness of recovered, weakly communicated implications of (18d), which account for the perceived creativity of the expression.[10]

The possibilities presented above are compatible with what I will call *the monosemy thesis*. On this thesis, the contextual proliferation of interpretations of lexical items does not give direct insight into their meaning. Many words encode unitary concepts (i.e. they open up singular conceptual addresses in memory) but may convey a rather different concept after being pragmatically manipulated. Although some of these *ad hoc* concepts,[11] i.e. contextual instantiations of the original conceptual entry, may become stabilised in memory, they do not have to be; indeed, standard methodological considerations of cognitive parsimony might militate in favour of keeping stored conceptual representations as unified as possible. On the monosemy story, the mapping between concepts and natural-language words is far from simple, since almost every word communicates a concept which departs - to a greater or lesser degree - from the concept which is encoded by that word.

The bias towards pragmatic explanations of the plasticity of lexical content does not, of course, settle the issue of adjudicating between the possibilities represented in (18a-d) (or other candidate analyses) for particular examples; on a more fundamental level, it does not resolve the problem of justifying the options in (18) as distinct types of lexical input

---

[10] In weak communication the speaker assumes only partial responsibility for the cognitive effects which the hearer recovers from the utterance, although she may encourage several lines of interpretation. In different terms, weak communication arises when the communicator's intention to marginally increase the hearer's awareness of a wide range of assumptions itself becomes only weakly manifest (see Sperber and Wilson, 1986/1995: 59-60, 1990, and next section).

[11] The term belongs to Barsalou (1983, 1987) and has been widely adopted within relevance theory; cf. Carston (1996), Sperber and Wilson (1997), Papafragou (1995).

to the pragmatic device. One of my aims in this book is to provide such a justification, and to separate the phenomena in (18) from either ambiguity or polysemy within a relevance-theoretic framework.

1.2.2 Relevance Theory and Utterance Comprehension

Humans live and communicate within a certain cognitive environment: a set of facts and assumptions which are manifest to them. Facts and assumptions are manifest to individuals to the extent that the individuals can mentally represent them and accept them as true with a high degree of confidence. On this picture, human communication is for the most part an attempt to make a set of facts or assumptions manifest, or more manifest, to an audience; what communicators aim to do (and competent communicators succeed in doing) is to affect others' thoughts in a partly predictable way by affecting their cognitive environment.

The means of human linguistic communication, utterances, are instances of ostensive behaviour. This means that they call attention to themselves in a particular way. Generally, ostensive stimuli provide access to two layers of information: they make manifest to the addressee the intention of the communicator to make it manifest that she intends to inform the addressee of something. Consider the situation in which I point to a van selling ice-cream on a very hot day. My gesture is an ostensive stimulus which makes manifest to you my intention to make it manifest that I intend to inform you of something (:that we can get ice-cream from the van). The point about ostensive stimuli is that they give the addressee adequate reason to suppose that they have some significance which non-ostensive stimuli lack. In other words, ostensive stimuli, by demanding the addressee's attention, warrant expectations of relevance in a way that non-ostensive stimuli do not.

A stimulus is relevant to an individual to the extent that it causes cognitive effects for that individual.[12] There are three broad types of cognitive effect: (a) the stimulus may interact with previously held assumptions to yield new implications (*contextual implications*, as Sperber and Wilson call them); (b) the stimulus may contradict an existing assumption and result in its elimination from the addressee's mental repertoire;

[12] For an utterance to be genuinely relevant to the individual (as opposed to merely seeming relevant), the cognitive effects must lead to a genuine increase in knowledge (i.e. must represent facts, as opposed to false assumptions) or contribute positively in some other way to the achievement of cognitive functions and goals. These are called *positive cognitive effects* - see the Postface to the second edition of *Relevance* (pp. 265-6) for discussion.

(c) the stimulus may offer support for an existing assumption and result in its strengthening in the individual's mental repertoire. Processing stimuli, however, requires some energy, or processing effort. This co-varies with factors such as the accessibility of the stimulus employed, its form (perceptual or linguistic complexity, length) and the size and accessibility of the chunk of assumptions in the addressee's mental repertoire which are required for the computation of cognitive effects. Relevance can, therefore, be seen as depending on a balance between effects and effort, or on the 'gains' from a certain stimulus and the cognitive energy it takes to retrieve those gains. The more weighty the gains are, the more relevant the stimulus becomes; the less effort is required to achieve those gains, the more the relevance of the stimulus is enhanced.

Utterances, like all ostensive stimuli, create a presumption of optimal relevance. This means that the addressee in a verbal exchange is entitled to assume that the conversational contribution which was directed at him was meant to offer him at least a satisfactory range of cognitive effects relative to the effort it would take to recover them, and to the rewards detachable by allocating that effort elsewhere. This is captured by the Communicative Principle of Relevance, which states:

> *Communicative Principle of Relevance:*
> Every act of ostensive communication (e.g. an utterance) communicates a presumption of its own optimal relevance.

Because they create a presumption of optimal relevance, utterances are expected to be at least relevant enough in terms of effect and effort, to be worth the addressee's attention, and moreover to be the most relevant, in terms of effect and effort, that the communicator is willing and able to produce. This can be formally stated as follows:

> *Presumption of optimal relevance*:[13]
> (a) The ostensive stimulus is relevant enough for it to be worth the addressee's effort to process it, and
> (b) The ostensive stimulus is the most relevant one compatible with the communicator's abilities and preferences.

During the process of understanding an utterance, the hearer is often faced with a choice between several interpretations, all of which are compatible with the linguistically

[13] This is the revised presumption of optimal relevance which is included in the second edition of *Relevance* (1995).

encoded content of the utterance. In the framework I have been describing, the hearer is justified in treating as the correct interpretation (and, therefore, the one which the speaker intended) the one which satisfies his expectation of relevance. This, in a nutshell, is an interpretation which the speaker might rationally have expected to be optimally relevant to the addressee.

On hearing an utterance, the addressee starts retrieving or constructing sets of mentally representable assumptions which might interact with his mental representation of the utterance. The set of mentally represented assumptions which the hearer may bring to bear on the interpretation process constitutes what Sperber and Wilson (1986/1995) call the *context* for utterance comprehension. Contextual assumptions will be retrieved or constructed according to their accessibility and will be processed together with the mental representation of the utterance to yield tentative interpretive hypotheses. These hypotheses will be tested against the hearer's expectation of relevance, i.e. his expectation (in normal circumstances) that the speaker will have aimed at, if not achieved, optimal relevance. The first hypothesis which is found to satisfy this expectation should be retained. A straightforward prediction of the model is that the hearer is entitled to assume that the first interpretation to satisfy his expectation of relevance will be the only one to satisfy his expectation of relevance - hence the hearer will be entitled to conclude that he has reached the correct interpretation of the utterance.

Expectations of relevance constrain a variety of sub-phases of verbal understanding. Recall that the extraction of the proposition expressed by an utterance from the schematic representation provided by the utterance's linguistically encoded content is a pragmatic task, hence amenable to relevance considerations. The hearer's expectation of relevance helps in a number of ways in fleshing out the underdetermined logical form of the utterance into the complete truth-evaluable representation which (it is assumed) the speaker intended to convey. Specific work has shown how disambiguation, reference assignment, enrichment and loosening are all guided by considerations of relevance (see, e.g., Wilson, 1992; Wilson and Sperber, 1993b; Matsui, 1995; Carston, 1993, 1996, 1997, 1998; Rouchota, 1992, 1994b; Blakemore, 1987; Blass, 1990, and the contributions in Rouchota and Jucker, 1998). As a brief illustration of how a relevance-based mechanism guides the utterance interpretation process, consider (19) as an exchange between two lawyers:

(19) Martin Gregg's case is difficult to handle.

I will focus on three aspects of the underdeterminacy in (19). Firstly, the hearer has to disambiguate *case*, which could mean (at least) 'lawsuit' or 'briefcase'. Secondly, he has to determine the interpretation of the possessive, which arguably encodes a very general relation between two constituents: in the above example, narrowing down the meaning of the possessive depends on the interpretation of *case*. Thirdly, *handle* may refer to a physical/manual manipulation or to a legal treatment. Suppose that Martin Gregg is another lawyer who has just undertaken the defence of an alleged paedophile. If this assumption is manifest to the interlocutors and accessible enough to the addressee, then it gives rise to the first interpretive hypothesis about the proposition expressed by (19). Very roughly, this first hypothesis is given in (20):

(20) The lawsuit which Martin Gregg is dealing with is difficult to handle.

Combined with highly accessible contextual assumptions about incidents of paedophilia, the legal difficulties presented, the sensitivity of the public, and so on, the hypothesised proposition expressed by (19) may give rise to an array of contextual implications which (it is assumed) the speaker intended to convey, i.e. an array of *implicatures*:

(21) a. Martin Gregg is very busy.
b. We won't see much of Martin Gregg in the next weeks.
c. Martin Gregg may lose this case.
d. Martin Gregg's professional reputation is in danger because of this case.

To the extent that the set in (21a-d) satisfies his expectations of relevance, and does so in a way the speaker could have foreseen, the hearer may conclude that he has reached the endpoint of comprehension and accept the hypotheses he has constructed as the intended interpretation of (19). Other interpretations remain linguistically possible, but are pragmatically dispreferred. It follows from the definition of optimal relevance that the hearer is not justified in going beyond the first acceptable interpretation and considering potential alternatives: this would incur additional processing effort on his part, which a competent speaker aiming at optimal relevance should manifestly have avoided.

The proposition expressed by (19) is (taken to be) an assumption which the speaker wanted to communicate, and is therefore the *explicature* of the utterance. Usually a speaker intends her utterance to make manifest (or more manifest) a variety of assumptions, which go well beyond the (main/base) explicature of the utterance: implicatures are one such type of assumption. There is another type of communicated assumption which I want to introduce. In producing (19), the speaker might have intended any of the following to be made manifest (or more manifest) to her interlocutor:

(22) a. The speaker is saying that the lawsuit which Martin Gregg is dealing with is difficult to handle.
b. The speaker believes that the lawsuit which Martin Gregg is dealing with is difficult to handle.
c. The speaker asserts that the lawsuit which Martin Gregg is dealing with is difficult to handle.
d. The speaker regrets that the lawsuit which Martin Gregg is dealing with is difficult to handle.

These further assumptions are produced by embedding the proposition expressed by (19) under speech-act descriptions or propositional-attitude reports. The resulting propositions (22a-d) are called *higher-level explicatures* within relevance theory (Sperber and Wilson, 1986/1996; Wilson and Sperber, 1993a). The (optional) recovery of these further assumptions is performed through a combination of decoding and inference, as in the case of the base explicature of the utterance. Often, the linguistic structure of the utterance suggests a particular attitude: the declarative form of (19) gives rise most naturally to a higher-level explicature such as (22a). On other occasions, tone of voice, facial expression, etc. may give a hint as to the speaker's attitude towards the base explicature of the utterance (cf. (22d)). Although part of the explicitly communicated content of the utterance, higher-level explicatures do not make a contribution to the truth conditions of the utterance: the truth conditional content of (19) is given in (20).

To conclude, I want to note that the partitioning of the information communicated by an utterance into explicit and implicit does not correspond to the semantics/pragmatics distinction. As the above example has shown, there are inferred (pragmatic) aspects of the explicitly communicated content of an utterance, both on the level of the base explicature and in higher-level explicatures. In the main part of the book, I will be concerned with delimiting the contribution of semantic and pragmatic devices to explicitly conveyed information in verbal communication.

2

THE SEMANTICS AND PRAGMATICS OF ENGLISH MODAL VERBS

> Castro visits Moscow and is taken on a tour by Brezhnev. First, they go for a drink and Castro praises the beer.
> 'Yes, it was provided by our good friends from Czechoslovakia'.
> Next, they go for a ride in a car and Castro admires the car.
> 'Yes, these cars are provided by our good friends from Czechoslovakia'.
> They drive to an exhibition of beautiful cut glass, which Castro greatly admires.
> 'Yes, this glass comes from our good friends in Czechoslovakia'.
> 'They must be very good friends', says Castro.
> 'Yes, they must', says Brezhnev.[1]

2.0 INTRODUCTORY REMARKS

In this chapter, I introduce and defend a semantic and pragmatic analysis of a sample of the English modal verbs. I start by discussing previous analyses of modality in English with an eye to explaining how a cluster of related meanings (epistemic, root, and other) is expressed by the same set of lexical items (2.1).

I go on to develop a unitary semantic approach to the English modals, treating them as incomplete or otherwise unspecified propositional operators (2.2). After defending the details of my semantic account, I show how the proposed semantics can give rise to the range of root interpretations modal verbs can receive in context (2.3). Epistemic interpretations require some further theoretical machinery, which I introduce and motivate in 2.4. Finally, I sketch the differences between natural-language interpretations of modal operators and their alethic/logical uses.

---

[1] From Lukes and Galnoor (1987: 52-3); cited in Smith (1989: 89).

2.1 Previous Analyses

2.1.1 The Ambiguity View

One possible approach to the systematic meaning multiplicity of the English modals is to assume massive lexical ambiguity; this is the view adopted by Palmer (1990). On this approach, each modal verb encodes a particular cluster of distinct modalities, as demonstrated by the following paraphrases:

(1) a. You may enter.
'You are allowed to enter'. (deontic)
b. The mountain may be climbed from other points on this tour.
'It is physically possible to climb the mountain... '. (neutral dynamic)
c. You may find this magazine interesting.
'You will perhaps find this magazine interesting'. (epistemic)

(2) a. You can come in.
'You are allowed to come in'. (deontic)
b. You can come in casual dress.
'The regulations allow you to come in casual dress'. (neutral dynamic)
c. Mary can tap-dance.
'Mary is able to tap-dance'. (subject-oriented dynamic/ability)

(3) a. You must leave immediately.
'You are obliged to leave immediately'. (deontic)
b. I must accept your resignation.
'The circumstances force me to accept your resignation'. (neutral dynamic)
c. You just must go around asking these indiscreet questions!
'It is a compulsion for you to go around asking... '. (subject-oriented dynamic)
d. You must be joking.
'You are certainly joking'. (epistemic)

(4) a. He should cut down on smoking.
'He is obliged to cut down on smoking'. (deontic)
b. The injection should be painless.
'I tentatively conclude that the injection will be painless'. (epistemic)

The ambiguity view faces a number of problems, both descriptive and explanatory. To begin with, there is an intuitive difference between the phenomenon of classic lexical ambiguity exemplified by *port* in (5) and the type of meaning multiplicity which characterises the modals; the latter seems to be motivated in a sense which goes well beyond simple ambiguity:

(5) This port is justly famous.

A number of arguments support this intuition. First, there is considerable semantic overlap between the postulated senses for the modals. For instance, a semantic distinction between subject-oriented and neutral dynamic modality is problematic, since the former reading is logically subsumed under the latter. This is obvious in the following examples:

(6) Mary can speak German.

(7) Mary can speak German at the meeting because everybody will understand.

Problems also appear in the broader distinction between dynamic and deontic modality. As (8) shows, dynamic *can* is capable of conveying a deontic meaning:

(8) You can be the first person to join our forces at such a young age.

Palmer concedes that deontic (permission) readings may be pragmatically derived from dynamic (possibility) ones; if generalised, this claim would open the way towards treating all deontic meanings of modals as products of pragmatic interpretation and not of semantic decoding processes.

As a second argument against the multiple ambiguity position, note that the proposed set of senses fails to capture the range of meanings which the modals contextually convey. Consider the following example (mentioned in Groefsema, 1995):

(9) You must come to dinner sometime.

None of the meanings provided above for *must* accounts for the normal interpretation of (9), which is something like (10):

(10) We would like you to come to dinner sometime.

Similarly, *can* may be used to communicate more than its proposed semantics would suggest (Walton, 1988: 103):

(11) a. You can clean the house for once.
b. Can you pass the salt?
c. Can I get you a drink?

The ambiguity view may attempt to incorporate such meanings in two ways: either by further complicating the semantics for modals (Leech, 1987), or by appeal to 'conventions of usage' or 'pragmatic extensions' (Palmer, 1986, 1990). In either case, descriptive adequacy is achieved at the expense of a truly explanatory account of the semantics and pragmatics of the modals.

Although the proponents of the ambiguity view may offer different accounts of modality, they are all subject to the above criticisms. For instance, Coates (1983) tries to accommodate the richness of possible interpretations of each modal verb (its 'indeterminacy') by representing its meaning as a fuzzy set (see also Leech and Coates, 1980). The indeterminacy is then attributed to the gradience between the semantic core and the periphery of each modal. In the case of *can*, for example, the continuum of meaning extends from the core of 'ability' to the periphery of 'possibility'; gradience is manifested in utterances like (12), where it is difficult to adjudicate between the two meanings:

(12) All we can do is rake up somebody like Piers Plowman who was a literary oddity.

In other modals (e.g. *must, may,* and *should*), deontic and dynamic aspects are grouped together under 'root' and are attributed distinct fuzzy set meanings from their epistemic counterparts. Coates argues that certain examples may be ambivalent as to whether they are intended to receive a root or epistemic interpretation on a given situation of utterance, as in (13); alternatively, root and epistemic interpretations may merge, as in (14):[2]

(13) He must understand that we mean business.

(14) a. Newcastle Brown is a jolly good beer.
b. Is it?

---

[2] The examples in (12)-(14) are taken from Coates (1983).

c. Well, it ought to be at that price.

Gradience, ambivalence and merger all contribute to the indeterminacy of modal interpretation.

Coates' account again is characterised by an excessive reliance on semantics to provide the whole array of meanings communicated by the modals; as a result, she constantly has to expand the semantic component so that it includes information about the degree of 'subjectivity' or 'strength' of the modality. Even then, the absence of a pragmatic component from her analysis makes it impossible to explain examples of the sort in (10)-(11). As shown by the use of the term 'indeterminacy', Coates is primarily worried by the inadequacy of semantic labels to yield an empirically satisfactory analysis of the modals: however, the problem would not arise if this indeterminacy were seen as resulting from the flexibility of pragmatic interpretation. A good piece of evidence against postulating phenomena such as gradience or merger is the fact that these phenomena do not actually have any traceable effect on the comprehension process: contrary to what one would expect on Coates' account, utterances of the sort in (12) are not more difficult to understand than (2c), an utterance conveying the 'core' meaning of *can.* This intuitive estimate seems to be supported by the fact that recent psycholinguistic research has consistently singled out and tested utterances of the type in (11) - especially (11b) - for speed of comprehension, thereby supporting the view that this class of utterances may cause some interesting processing complications; no similar hypothesis has ever been considered for Coates' gradience or merger examples (for experimental results, see Gibbs, 1994: 88-91).

To summarise: although the fundamental point of the ambiguity-based approach is the rigid distinction between the epistemic and various non-epistemic 'meanings' of the modals, both Palmer and Coates are forced to recognise a wide range of intermediate cases, where for a variety of reasons the proposed semantic distinctions prove inert, indistinguishable or insufficient. These cases, however, may be viewed as a threat to the overall validity of the ambiguity position. A more viable alternative, which I have already hinted at, would explore the view that traditional categories of modality are effectively by-products of the comprehension process rather than stable and basic semantic information guiding pragmatic interpretation. A more satisfactory account of the modals should then offer a more parsimonious semantics and a richer pragmatics; the latter should be able to deal with issues of politeness, style or illocutionary force. Moreover, this account should explain why epistemic and root meanings can be conveyed by the same class of lexical items.

This last issue brings us to a modification of the ambiguity view, put forth by Sweetser (1990).

2.1.2 The Polysemy View

Sweetser (1990) places her discussion of modality within a more general study of polysemy in natural language. Adopting a Cognitive Linguistic framework, she claims that polysemy is often motivated by a metaphorical mapping from the concrete, external world of socio-physical experience to the abstract, internal world of reasoning and of mental processes in general. She argues that modal verbs display a similar, motivated polysemy, thus rejecting the traditional view that they are ambiguous between unrelated senses.

Sweetser uses as a basis for the semantics of the modals Talmy's (1988) notion of 'force dynamics'. Root modals are taken to encode force-dynamic notions in the external world. For instance, *may* encodes the existence of a potential but absent barrier, *must* a positive compulsion, and *can* either a positive ability on the part of the doer, or some potential force/energy. These notions are extended metaphorically into the internal, 'mental' domain and give rise to epistemic meanings: *may* and *must* thus come to denote barriers or forces operating in the domain of reasoning. To illustrate: (15')-(18') give pragmatically enriched paraphrases of the semantic content of the utterances in (15)-(18) respectively; (15) and (16) are examples of root modality, whereas (17) and (18) exemplify epistemic uses of modals:

(15) You may spend this sum any way you wish.

(15') You are not barred (by some or other authority) from spending this sum any way you wish.

(16) You must be back by midnight. (Our parents said so).

(16') The direct force (of our parents' authority) compels you to be back by midnight.

(17) The butler may have committed the murder in the meantime.

(17') I am not barred by my premises from the conclusion that the butler has committed the murder in the meantime.

(18) The guests must have had a really good time.

(18') The available evidence compels me to the conclusion that the guests had a really good time.

On this account, the metaphorical mapping between root and epistemic senses is part of the semantics of English, i.e. it has become conventionalised. For individual occurrences of modal items, it is the task of pragmatic interpretation processes to decide which of the two domains (root or epistemic) is the intended one, i.e. to resolve the structured polysemy in the modal semantics.[3] Sweetser concludes with some arguments about how an account based on 'modality in two worlds' could explain the acquisitional and historical priority of the root over the epistemic meanings of the modals.[4]

Although Sweetser's approach correctly moves in the direction of supplying motivation for the systematic relation between root and epistemic uses of modal expressions, there are some points to be made about the specifics of the analysis. Notice, firstly, that even if one adopts the idea of a metaphorical mapping among modal concepts, this mapping will come out as very different from other examples of metaphorical mapping which have been claimed to motivate lexical polysemy. Consider the case of perception terms (e.g. *see*, *view*, etc.), which have displayed a cross-linguistic tendency to develop meanings related to mental processes. According to Sweetser (1990: 23ff.), this can be explained in terms of a metaphorical construal of the internal world delivered by reasoning on the basis of the external world delivered by perception. The semantics of perception terms thus includes a metaphorical mapping which relates two independent and distinct senses. In the case of modals, however, the senses allegedly linked through metaphor are not so distinct, as a range of indeterminate examples in the previous section has demonstrated. Consequently, even if we grant that metaphorical mappings exist in the case of perception verbs, the parallel between these and modals is not straightforward.

A second problem for the proposal based on a metaphorical extension of modal meanings is that its application is constrained in various ways. An obvious case is positive *can*, which is not normally used epistemically. There is no motivation for this fact in Sweetser's account, as she herself acknowledges (1990: 154). Sweetser considers this as an argument against a unitary semantic approach to the modals: 'It is not the case (as we might expect if the modals were simply monosemous) that all root modals must/can have epistemic uses - this is neither historically true for the English modals nor a cross-linguistic universal' (1990: 68). It is not necessary, however, for a monosemous account to make such

---

[3] See also Langacker (1991: 273ff.) for a polysemy-based approach to modals which builds on Sweetser's analysis.

[4] Sweetser goes on to tentatively propose a further mapping of force dynamics in modal verbs, this time into the speech-act domain. I postpone the discussion of that part of her proposal until chapter 3.

an assumption; on the contrary, a single semantics for the modals could leave room for a pragmatic explanation of the gaps in their distribution (see 2.3.1).

There is also something to be said on the historical basis of Sweetser's arguments, namely the precedence of the root over the epistemic meanings.[5] In the first place, there seems to be some evidence against the priority of root meanings. Goossens (1982) claims that epistemic uses of the modals existed in Old English, though they had a more restricted distribution than root ones. For instance, *magan* ('may') had already begun to express epistemic possibility, while other root modals like *willan* ('will') and *sculan* ('shall') were occasionally used to express epistemically coloured predictions. Although he claims that there was no modal counterpart for the present-day English epistemic *must*, Tanaka (1990) argues that *sculan* took up the position corresponding to the latter.

In the second place, semantic change in the modals did not involve only their root and epistemic meanings, but also took place between two root senses. For example, Shepherd (1982) reports the semantic development of *can* from initially expressing intellectual capacity in Old English to expressing general capacity and later possibility or permission. This, according to Tanaka (1990), was partly due to *can* taking over parts of the meaning of Old English *may*, which consequently changed from expressing general ability to expressing permission and possibility. It seems, therefore, that the historical development of the meanings of the modals cannot be explained in terms of a simple metaphorical mapping along the lines proposed by Sweetser. In any case, the extent to which diachronic evidence is relevant for a synchronic analysis of linguistic competence is a fairly controversial issue. Even if root meanings were the first to appear, the semantics of the English modals may well have developed towards a unitary meaning. As for the development of epistemic interpretations in items which initially encoded root concepts, there may be alternative explanations; I will briefly explore one such alternative in the appendix to chapter 3.

Polysemy-based accounts of modal expressions in English and other languages have also been advocated by a variety of scholars, especially within the grammaticalisation literature (Bybee, 1988a, b; Bybee and Fleischman, 1995b; Bybee and Pagliuca, 1985; Bybee *et al.*, 1994; Heine *et al.*, 1991; Traugott 1982, 1988, 1989, 1995;Traugott and König, 1991; cf. also Nuyts, 1993; Chung and Timberlake, 1985; Lyons, 1977; Halliday, 1970); I will not examine these proposals in any greater detail here (although see section 3.2 for relevant comments). The conclusion from the discussion so far seems to be that neither the ambiguity nor the 'motivated polysemy' approach can give a plausible semantics for the English modals.

---

[5] My discussion of the historical arguments draws heavily on Groefsema (1995: 59).

2.1.3 The Monosemy View

An obvious way of avoiding the problems of ambiguity/polysemy is to adopt a common core for the meaning of each modal, and to use it as a basis for deriving the vast range of possible interpretations which the modals may contextually receive. Earlier monosemous approaches to the English modals include Ehrman (1966), Wertheimer (1972), Tregidgo (1982), Perkins (1983), and Haegeman (1983). Here, I will review in some detail two proposals which have adopted a precise, formally statable semantics for modals, and a sound division of labour between semantic and pragmatic aspects of the comprehension of modality.

2.1.3.1 Kratzer's Account. Kratzer (1977, 1981a, 1991) provides an analysis of modality couched within a possible-worlds framework. In possible world semantics, utterances of sentences are taken to express propositions. A proposition is identified with the set of possible worlds in which it is true. The following definitions can serve as preliminaries to the discussion (assuming W is a set of possible worlds):

> A proposition p is true in a world $w \in W$ iff $w \in p$. Otherwise, p is false in w.
>
> A proposition p follows from a set of propositions A iff p is true in all worlds of W in which all propositions of A are true.
>
> A proposition p is compatible with a set of propositions A iff $A \cup \{p\}$ is consistent.

To explain how modality works in such a framework, Kratzer introduces three factors which jointly underlie modal operators: the *modal relation*, the *modal base* and the *ordering source*. The modal relation is basically the relation of compatibility or logical consequence (these relations underlie the notions of possibility and necessity respectively in a way to be made precise soon). The modal base (or *conversational background*) involves a set of assumptions against which the modal relation can be understood. In the following examples, the paraphrases with the phrase *in view of* give the preferred modal base for the interpretation of the utterance:

(19) a. James must be at work; he doesn't answer the phone.
'In view of what is known, James must be at work'.
b. Soldiers must do their duty.

'In view of what is commanded, soldiers must do their duty'.

c. I must buy a bicycle.

'In view of my goals, I must buy a bicycle'.

(19a) involves an epistemic modal base, (19b) a deontic modal base and (19c) a teleological modal base. Kratzer (1981a) mentions a variety of modal bases, which further include categories such as *realistic, stereotypical* or *buletic* (related to wishes).

Formally, modal bases are treated as functions from possible worlds to sets of propositions of a certain kind. Take the phrase *what is known*. What is known is different from one possible world to another. And what is known in any given possible world is a set of propositions. In (19a), the denotation of the phrase *in view of what we know* is a function which assigns to every possible world the set of propositions K such that what is known provides K in that world. Likewise for the phrase *what is commanded* in (19b): what is commanded differs from one possible world to the next. The denotation of the phrase is a function which assigns to every possible world the set of propositions C such that what is commanded provides C in that world. Generally, if W is a set of possible worlds, modal bases are functions which assign to every member of W a subset of the power set of W.

A first approximation to the meaning of modal words is to say that they express possibility or necessity with respect to different types of modal base. *Must*, for instance, as shown in (19), is capable of expressing necessity with respect to a variety of modal bases. Assuming that a is any sentence, f any modal base, and $[[a]]^f$ stands for the proposition expressed by a with respect to f, a preliminary definition of the meaning of *must* is given in (20) (Kratzer, 1991: 641):

(20) $[[\text{must a}]]^f = \{w \in W: [[a]]^f \text{ follows from } f(w)\}$

According to this definition, the proposition expressed by the utterance in (19a) is true in a world w if and only if it follows from what is known in w that James is at work. Similarly, in (19b) the proposition expressed is true in world w if and only if it follows from what is commanded in w that soldiers do their duty.

It is possible to give an alternative formulation of the definition in (20) in terms of accessibility relations familiar from standard modal logic (see, e.g., Hughes and Cresswell, 1968). An accessibility relation is a binary relation on the set of all possible worlds. Each conversational background comes with an accessibility relation. A world w is epistemically accessible from a world w if and only if w' is compatible with what is known in w. A world w' is deontically accessible from a world w if and only if w' is compatible with everything the law provides in w. Any conversational background f uniquely determines an accessibility relation R_f (Kratzer, 1991: 642):

(21) For all w, w' $\in$ W: wR_fw' iff w' $\in \cap f(w)$

(20) can be recast in terms of accessibility relations as follows:

(22) $[[\text{must a}]]^f = \{w \in W: [[a]]^f \text{ follows from } f(w)\}$

There are reasons for extending the present account (which Kratzer calls 'the standard analysis'). Notice that not all worlds contained in a modal base are taken into account in interpreting an utterance which contains a modal. For instance, in (19a) the hearer does not consider some epistemically accessible worlds in which James is suffering from an anti-social crisis, or has suddenly gone mute (and therefore fails to answer the phone). Why should that be? A first approximation is that worlds containing these states of affairs are further removed from the actual world and not easily accessible from it; they represent 'wilder' possibilities, which need not be taken into account. More precisely, they are removed from what we take to be the ideal or normal course of events. Therefore, we need an *ordering source* which will rate worlds in terms of similarity to an ideal and specify the minimal degree of 'distance' from the ideal a world should satisfy.

Ordering sources are another kind of conversational background involved in modal reasoning. Consider (19a): one type of conversational background (the epistemic modal base) determines for every world the set of worlds which are epistemically accessible from it. Another kind of conversational background (a stereotypical ordering source) induces an ordering on the set of worlds accessible from each world. The ordering excludes those worlds which are further removed from the normal course of events. The interaction of the two backgrounds ensures the correct interpretation for the utterance.[6] The idea of a partial ordering of possible worlds in terms of similarity to some norm (originally due to David Lewis[7]) is made concrete in Kratzer (1991: 644) in the following way:

> A world w is at least as close to the ideal represented by [a set of propositions] A as a world w' iff all propositions of A which are true in w' are true in w as well.

---

[6] In Kratzer's model, either one of the two conversational backgrounds can be empty, thereby placing no restrictions on the type of possible world the modal ranges over.

[7] Kratzer cites Lewis (1981); cf. also Lewis (1973a, 1986).

In the revised analysis, modality is thus relativised with respect to two kinds of background information: modal bases and ordering sources. Modal notions expressed by modal terms in natural language can now be formally defined in a more satisfactory way; *must* is taken to encode necessity, where necessity is construed as follows:

> A proposition p is a necessity in a world w with respect to a modal base f and an ordering source g iff the following condition is satisfied:
> For all $u \in \cap f(w)$ there is a $v \in \cap f(w)$ such that $v \leq_{g(w)} u$, and
> for all $z \in \cap f(w)$: if $z \leq_{g(w)} v$, then $z \in p$.

According to this definition, a proposition is a necessity if and only if it is true in all accessible worlds which come closest to the ideal established by the ordering source. The propositions expressed by the utterances in (19a) and (19b) can be paraphrased according to the new definition:

(19) a. James must be at work.
b. Soldiers must do their duty.

(19') a. In all those epistemically accessible words which come closest to the ideal, James is at work.
b. In all those deontically accessible worlds which come closest to the ideal, soldiers do their duty.

To mention one more example, *might* in this system encodes possibility. A proposition is a possibility if and only if its negation is not a necessity, or more precisely:

> A proposition is a possibility in a world w with respect to a modal base f and an ordering source g iff ~p is not a necessity in w with respect to f and g.

The proposition expressed by the utterance in (23) is paraphrased in (23'):

(23) It might rain.

(23') In all those epistemically accessible worlds which come closest to the ideal, it rains.

Different modals encode distinct clusters of modal relations, modal bases and ordering sources (see Kratzer, 1981a, 1991, for examples from German and English). The sort of modal base and ordering source that is appropriate for the interpretation of individual modal words is generally determined pragmatically (*modulo* certain lexically specified restrictions on admissible conversational backgrounds). Hence modal expressions are treated as context-dependent and vague, rather than lexically ambiguous items.

The picture I have outlined is extremely appealing as a theoretical model of the semantics of modality. Kratzer's proposal is particularly successful in assigning to modal expressions a weak semantics, which together with additional contextual considerations yields epistemic, deontic, etc. interpretations. My comments in what follows are meant as an attempt to build on Kratzer's ideas while paying more detailed attention to pragmatic contributions to the derivation of modal interpretations.

It is worth noting from the outset that the possible-worlds model is not meant as a psychologically plausible model of how speakers represent and handle alternative possibilities. In fact, the model taken at face value is hard to reconcile with experimental findings, which have consistently attested that human subjects have difficulty with the systematic mental manipulation of even a few alternatives (Johnson-Laird, 1982). The psychological reality of possible worlds is the subject of a long and heated debate which falls outside the scope of the present discussion (for some criticisms, see Smith, 1983; Smith and Smith, 1988). One way of preserving the model while dealing with the demands for psychological plausibility would be to suggest that worlds do not correspond to full-fledged representations of states of affairs but to partial specifications of them, or *situations*; another would be to assume that sets of possible worlds are not individually represented but subsumed under a single description.[8] In any case, the theory as outlined above should be taken to capture some facts about the speaker's semantic competence while remaining neutral as to how this competence is to be cognitively represented (see similar suggestions in Stalnaker, 1986: 120-1).

Even though a possible-worlds analysis is not designed to fit with a cognitively informed pragmatic theory, it seems that two of the three components of modality in Kratzer's proposal, namely the conversational background and the ordering source, involve non-linguistic knowledge and consequently belong to pragmatics. Kratzer herself makes suggestive and repeated references to the role of contextual factors in the overall

---

[8] Kratzer (1980), arguing against the view that possible worlds are incompatible with finite brains, shows that to know a set does not necessarily require a discrete mental representation of all its members; therefore, a given set of possible worlds need not require an individuated mental representation but may be stored under a single description (see also Partee 1989: 117ff., 1977). A similar proposal for the mental representation of time was put forth by Hans Kamp and is discussed in Johnson-Laird (1982).

comprehension of modals. Although promising, the way the conversational background and the ordering source are treated in her model requires further elaboration.

Firstly, one might get the idea from this exposition that there is a fixed inventory of conversational backgrounds favouring one modal interpretation over others. This idea, however, presupposes a rather inflexible conception of the role of context in comprehension. In this sense, it reflects earlier views in the literature on communication, according to which context was a determined/given chunk of information fixed independently of the utterance (see, e.g., Brown and Yule, 1983). As Sperber and Wilson (1986/1995) have shown, however, constructing the context for understanding an utterance is part of the interpretation process, constrained by general pragmatic principles.

Secondly, one can express serious doubts as to the usefulness and feasibility of an abstract (and absolute) ordering of possible worlds in terms of similarity to an ideal/normative standard. The point is that, whatever the ontological status of possible worlds, their similarity ratings are a cognitive and thus subjective matter. Even as ardent an adherent of modal realism as Lewis rejects the possibility of objectively measuring similarity among worlds (e.g. by mathematical methods). He admits (Lewis, 1973a: 91):[9]

> Overall similarity consists of innumerable similarities and differences in innumerable respects of comparison, balanced against each other, according to the relative importances we attach to these respects of comparison. Insofar as these relative importances differ from one person to another, or differ from one occasion to another, or are indeterminate even for a single person on a single occasion, so far is comparative similarity indeterminate.

In a more revealing excerpt, Lewis points out that our perception of similarity makes little use of the vast inventory permitted by logic: interlocutors expect each other to remain within a relatively limited range of inter-world similarity, and it is natural to have vocabulary conventionally reserved for use within this range (1973a: 94). In what follows, I will take seriously the idea that humans obey cognitive and communicative constraints in their assessments of similarity (and, in particular, similarity among worlds); indeed, in a later section, I will show that the limited vagueness of similarity which Lewis noted can be attributed to relevance-based limitations in construing plausible alternatives to the actual world.[10]

---

[9] See also the well-known puzzles of Goodman (1970).

[10] Interestingly, Kratzer (1989) in her analysis of counterfactuals points out - without further comment - that, in determining the set of propositions relevant for their truth, only 'humanly graspable' propositions should

Finally, there is something to be said about the status of the root-epistemic distinction in the above account. Kratzer argues that there are clear intuitive differences between modals with root ('circumstantial') and epistemic interpretations. The differences, however, are not semantic but pragmatic, and are due to the existence of two distinct major classes of modal bases (Kratzer, 1991: 646):

> Epistemic and circumstantial modal bases are both *realistic modal bases.* That is, both kinds of conversational backgrounds assign to every possible world a set of propositions which are true in that world. Yet circumstantial and epistemic conversational backgrounds involve different kinds of facts. In using an epistemic modal, we are interested in what else may or must be the case in our world given all the evidence available. Using a circumstantial modal, we are interested in the necessities implied by or the possibilities opened up by certain sorts of facts.

As in all formal semantic accounts, epistemic modality is treated as involving 'what is known' – where *know* is a factive predicate. On these analyses, what is known is a subset of what is the case. A cognitive approach to modality would start from different premises. From a cognitive perspective, epistemic modality is not concerned with what is known/what the evidence is in general, but rather with what the evidence is *from the point of view of a specific individual* (the speaker). Furthermore, on the internalist picture, epistemic modality does not involve knowledge in the strong sense but belief – and *believe*, unlike *know*, is not a factive predicate.[11] I agree with Kratzer that there is a (pragmatic) difference between root and epistemic interpretations of modal expressions. However, the difference seems to lie in the fact that epistemic modality involves the speaker's mental representation of reality and the evaluation of the (mentally modelled) evidence for that representation, whereas root modality involves relations among circumstances. One of my objectives in the pages to follow is to substantiate and motivate this way of drawing the root-epistemic distinction.

Kratzer explicitly states that she does not want 'to get into the mess of context theories' (1977: 343), and intentionally limits herself to a mere sketch of pragmatic aspects of meanings communicated by modal expressions. Obviously, an adequate pragmatic framework including a theory of context selection could profitably complement Kratzer's proposal, while maintaining the important insights of her semantic analysis. More

---

be considered; cognitive factors are thus again indirectly introduced in the formal analysis (cf. Kratzer, 1981b).

[11] These issues are connected to the difference between 'subjective' and 'objective' epistemic modality. For relevant discussion, see section 2.4.3.

specifically, such an account could show how the concepts encoded by modal expressions undergo inferential manipulation in order to yield the complete propositional representation of the modal utterance; this representation would then receive a truth-theoretic analysis of the sort proposed by Kratzer. I take up and expand on this idea in constructing a framework for modality in later parts of this chapter.

2.1.3.2 Groefsema's Account. Groefsema's (1995) analysis, cast in a relevance-theoretic framework, sets out to bridge the gap between the unitary semantics and the diverse pragmatics of modals.[12] Groefsema proposes the following semantics for *can*, *may*, *must* and *should* (where p is the proposition expressed by the rest of the utterance):

> *can*
> p is compatible with the set of all propositions which have a bearing on p
>
> *may*
> there is at least some set of propositions such that p is compatible with it
>
> *must*
> p is entailed by the set of all propositions which have a bearing on p
>
> *should*
> there is at least some set of propositions such that p is entailed by it

Bearing is defined as a formal relation between propositions (Groefsema, 1995: 62):

> A proposition P positively has a bearing on proposition Q iff
> a) [Q] or [~Q] follows from P, or
> b) [Q] or [~Q] follows from P and the minimal set of propositions X which, together with P will yield [Q] or [~Q] (condition: [Q] or [~Q] doesn't follow from X alone).
> A proposition P negatively has a bearing on proposition Q iff
> a) [Q] or [~Q] follows from ~P, or

---

[12] Smith (1989) contains an early proposal for treating *must* as monosemous within a relevance-theoretic framework. Other relevance-theoretic accounts of the English modals include Walton (1988) and Klinge (1993). I concentrate on Groefsema's proposal, since it is the most fully and articulately developed.

b) [Q] or [~Q] follows from ~P and the minimal set of propositions X which, together with [~P] will yield [Q] or [~Q] (condition: [Q] or [~Q] doesn't follow from X alone).

Groefsema goes on to point out that 'in informal terms, what the notion of bearing does is to focus the addressee's attention on all the *evidence* (of whatever nature, epistemic or otherwise) for the proposition expressed by the rest of the utterance' (ibid. pp. 62-3, her emphasis). Depending on the sort of evidence available and its relation to the embedded proposition p, an utterance containing a modal verb will be interpreted epistemically, deontically, etc. The interpretation process will always be constrained by the criterion of consistency with the principle of relevance.

Groefsema's account undoubtedly moves the study of modals forward by adopting a well-worked out division of labour between semantic content and pragmatic interpretation. I agree with a lot of individual points in her analysis and find the case she makes for bare modal semantics quite compelling. However, there are several remarks to be made about the details of her account. To start with, the crucial notion of bearing seems to be rather artificial when one considers how modals are actually used in communication. Take an epistemic interpretation such as that in (24), where the notion of evidential support intuitively makes more sense:

(24) Sue and John must be a couple.

On anyone's account, epistemic *must* conveys that the embedded proposition, say p, follows from a set of background propositions/evidence available to the speaker. Groefsema allows for two ways in which this can happen: evidence can either positively or negatively have a bearing on p. In the former case, the speaker must have had in mind something like the following: 'p is entailed by the set of all propositions which positively have a bearing on p' – where this set includes either

i) propositions which entail p or ~p, or

ii) propositions which are members of the minimal set of propositions which jointly entail p or ~p.

On (i), (21) turns out to be either a tautology or a contradiction, and it definitely is neither; (ii), to the extent that it is intelligible, leads to the same results as (i). This account has not given us a plausible semantics to use as a starting point in order to pragmatically flesh out what (24) communicates.

What about the possibility of the evidence negatively bearing on p? In this case, the speaker in (24) must have reasoned as follows: 'p is entailed by the set of all propositions which negatively have a bearing on p' - where this set includes either

i) propositions whose negation entails p or ~p, or

ii) propositions whose negation is a member of the minimal set of propositions which jointly entail p or ~ p.

The problem with this is that, in drawing a conclusion, we rarely - if ever - consider the *negation* of the available evidence; we may consider a varying *range* of evidence, but to depart from what we know to be the case would have little foreseeable cognitive gain. In (24), it does not seem possible to construct a context in which the speaker could plausibly reason on the basis of propositions with a negative bearing on the embedded assumption: if, for instance, the speaker's evidence in (24) consists (roughly) of the propositions in (25), it is difficult to see what role their negation could play in producing (24):

(25) a. Sue and John walk hand in hand.
b. A boy and a girl do not walk hand in hand unless they are a couple.

Apart from issues of production, there is also the issue of comprehension of negative evidence: although the propositions which have a bearing on p are not always retrieved by the hearer, Groefsema allows this to happen if it is crucial for establishing the relevance of an utterance containing a modal verb (1995: 69). However, there do not seem to be any occasions on which hearers work out what the negative evidence for the conclusion 'Sue and John are a couple' is or could be, let alone believe that the speaker would wish to make such evidence manifest by uttering (24). Interestingly, nowhere in the vast literature on modality and evidentials is there any mention, as far as I know, of the role of negative evidence, and this squares well with both intuition and empirical facts about the comprehension of modal expressions. Groefsema herself in practice abandons her technical definition by ignoring negative bearing; in analysing actual examples, she makes exclusive use of positive bearing in conjunction with a pretheoretical notion of evidence. This is no coincidence, since both of these notions hook onto something intuitively closer to modal meanings than their formally developed but somehow artificial counterparts which she previously proposed.

I have a number of other remarks concerning Groefsema's account. For instance, as already mentioned, Groefsema believes that only on some occasions is the hearer justified in looking for the specific propositions that have a bearing on p. However, it seems that, in order to understand a modal utterance, the hearer has to recover at least a broad specification of the *sort* of propositions that have a bearing on p, so that he can retrieve one or other type of modal interpretation. As an example of a case where the addressee does not have to work out what the evidence is, Groefsema offers the following:

(26) Ann must be at Heathrow by now.

She comments that the hearer can conclude from (26) that all the evidence available to the communicator entails that Ann is at Heathrow (at a given time): if the hearer trusts the communicator to know enough about Ann's movements, he will not have to work out what this available evidence is but can simply infer that Ann is at Heathrow (ibid. p. 69). It seems, though, that even in this example, the hearer had to compute the type of evidence (i.e. epistemic), even though not the individual assumptions, which the communicator entertained and used as grounds for her utterance.

Groefsema makes a further comment on the interpretation of (26). Following Walton (1988), she remarks: 'if the addressee does not believe that the communicator actually knows anything about Ann's movements, s/he will interpret [26] not as a description of an actual state of affairs, but rather as a description of a desirable state of affairs' (1995: 69) - whereby (26) yields (26'):

(26') The speaker desires that $_q\{[_p$Ann be at Heathrow by time *t*] is entailed by the set of all propositions which have a bearing on p$\}$.

(26') is distinctly odd: how can one desire that a certain entailment relation exist? Further, if something is desirable, shouldn't it be that Ann be at Heathrow (at a specific time), rather than what q in (26') denotes? More to the point: it is simpler to assume that the speaker in (26) is making a wild guess based on scarce evidence. This leaves us with a normal epistemic interpretation, albeit affected by manifest assumptions about the speaker's access to information about Ann's whereabouts.

In the following section, I want to present an alternative semantic analysis. Although it has similarities to both Kratzer's and Groefsema's accounts, this analysis departs in significant respects from previous approaches to modality.

2.2 A Semantic Proposal

2.2.1 Modality and Quantification

I am going to adopt an assumption already familiar from other monosemous approaches: modals are context-dependent expressions, in that their linguistic semantics radically underdetermines the overall meaning they communicate. I assume, partly agreeing with Kratzer, that the semantic content of modals consists of two components: a logical relation R (basically: entailment or compatibility), and a domain D of propositions. Roughly, then, what the modals are used to convey is that a certain proposition p bears a certain logical relation R to the set of propositions in a domain D, or, schematically:

(27) R (D, p)

The structure in (27) is an instance of a general tripartite structure which has been proposed for a number of quantificational devices in natural language, such as conditionals, *when*-clauses, quantificational determiners (*everyone, all*) or adverbs (*always, generally, often* - see Lewis, 1975; Heim, 1982; Farkas, 1981; Farkas and Sugioka, 1983; Schubert and Pelletier, 1989; Partee, 1989, 1995a; Roberts, 1989, 1995). This tripartite structure has the following form (Krifka *et al.*, 1995):

(28) OPERATOR (Restrictor, Matrix)

The operator takes scope over the proposition in the matrix and relates it to another proposition (the restrictor). In the case of modals, the operator is the logical relation of entailment or compatibility, the matrix is the embedded proposition p and the restrictor is the domain of propositions which the matrix is being placed in relation to; it is the restrictor that is responsible for the different types of modal concepts which a modal expression is capable of expressing in different contexts. As is often the case with tripartite quantificational structures, the restrictor in modals may be either linguistically indicated in some way, as in (29), or pragmatically inferred, as in (30):

(29) a. In view of the political situation, you must leave the country.
b. If you want to avoid a scandal, you must leave the country.
c. In case you plan to save yourself, you must leave the country.

(30) You must leave the country.

Regardless of the (phonologically realised or null) form of the restrictor in the surface structure of the modal utterance, the semantic content of a modal item may make reference to the possible restrictors (i.e. domains of propositions) it admits; alternatively, the semantic entry may remain silent as to the admissible restrictors, thereby leaving their specification entirely up to pragmatic processing. It will turn out that both cross-linguistic differences and historical changes in the semantics of modal expressions mostly turn on facts about/developments in the type of admissible restrictors. For the moment, however, I want to explore the differences within a subset of the English modals as to the modal relations and restrictors specified by the semantics of each verb.

My analysis of the semantics of modals will proceed in three steps. Firstly, I will develop the notion of domains of propositions which may serve as restrictors for modal operators. Secondly, I will introduce semantic analyses for a sample of the English modal verbs. Thirdly, I will elaborate on the specific types of context-dependence exhibited by modal verbs by motivating both pragmatic saturation and pragmatic enrichment analyses to deal with individual modals. I will consider each step in turn.[13]

2.2.2 Modal Restrictors

Following partly Sperber and Wilson (1986/1995; Sperber, 1997), I assume that any given proposition can be entertained and stored in memory in several different ways. Firstly, and most obviously, a proposition can be entertained as a truth-conditional description of a state of affairs in the actual world - in different terms, a factual assumption (a case of what relevance theory calls the *descriptive* use of propositions - see Sperber and Wilson, 1986/1995). Factual assumptions are the means whereby we represent reality to ourselves. They come in two main varieties: on the one hand, there are factual propositions describing wide-ranging, empirical generalisations about classes of objects and events; on the other hand, there are specific factual propositions concerning instances of events or particular individuals at given temporal and spatial locations. Factual assumptions are the default (or base) type of assumption for the purposes of communication, since they form a rich and highly accessible contextual background against which ostensive stimuli are processed. Propositions describing the actual world can thus be considered to belong to a single domain - the *factual* domain.

[13] Versions/parts of the semantic analysis of the modals I propose here appear in Papafragou (1998a, b, d, f).

Other domains of propositions include more constrained systems of laws, regulations or rules; I will call them *regulatory* domains. Regulatory domains include legal rulings, social regulations, religious rules, chess rules, etc.

Similar to these are domains in which propositions are handled as descriptions of states of affairs in ideal, or stereotypical worlds; at least as far as a sub-case of ideal-centered domains, the domain of moral beliefs, is concerned, there has been some evidence that it is stored and handled by a purpose-specific internal module (Premack and Premack, 1994).

Yet other types of domain may involve propositions which are handled as descriptions of states of affairs in worlds desirable from someone or other's point of view. Desirability is a three-place predicate: an individual can entertain an assumption as a description of a state of affairs in a world desirable from that individual's or someone else's point of view (Wilson and Sperber, 1988a).

Finally, propositions can be entertained and stored as abstract representations (i.e. hypotheses), or abstract representations of representations (where the initial representation may or may not be attributed to some source). These are examples of what relevance theory calls the *interpretive* use of propositions, and will come out as a separate domain of propositions (for further elaboration of this last type, see section 2.4).

On this picture, then, propositions are organised in domains. Logical relations such as entailment or compatibility apply only among propositions within a given domain.[14] I intend the types of domains I have discussed to be neither exhaustive nor mutually exclusive. It is fairly obvious, for instance, that normative and ideal-centered domains will overlap. Moreover, this discussion is by no means intended to imply that propositional domains are rigid and pre-constructed mental structures (cf. the comments about Kratzer's conversational backgrounds); I rather intend them as a sort of file-based organisation of our belief-desire system, which has a rich enough internal structure to be updated and expanded through the formation of novel domains. For purposes of constructing a framework for modality, domains of propositions establish a notional space which fulfils a two-fold purpose: in the first place, it offers the means for pragmatically restricting modal relations. In the second place, it provides a conceptual pool for grammaticalisation processes to draw on. As I have already suggested in my general discussion of restrictors, individual modal expressions will come out as permitting different kinds of domains of propositions as restrictors. Consequently, domains of propositions will prove a useful tool for describing

[14] The idea that propositions are organised in domains has been around for some time; its construals differ depending on the content which different proposals have ascribed to the term 'proposition'. For relevant discussion I refer the reader to Kuroda's (1982) indexed predicate calculus, Fauconnier's (1985) mental spaces and Recanati's (1995) domains of discourse (cf. also Kratzer's, 1981a conversational backgrounds, with which my own proposal is most closely related).

and comparing modal items in the same or different languages, or for tracing historical developments in modal meanings. Even if it turns out that it is not entirely accurate to treat domains of propositions as underlying mental structures, or (when grammaticalised) as contributions to semantic meaning, they can nevertheless be considered as a useful way of thinking about modal structure.

2.2.3 Semantics for Modal Operators

I now want to introduce semantics for a subset of the English modals. I propose that the information grammatically assigned to *can, may, must* and *should* is the following:

Can: p is compatible with D_{factual}

May: p is compatible with $D_{\text{unspecified}}$

Must: p is entailed by $D_{\text{unspecified}}$

Should: p is entailed by $D_{\text{normative}}$

(p: the embedded proposition; D: set of propositions in a domain)

On this proposal, *may* and *must* turn out to be semantically more general than *can* and *should* respectively: the verbs in the former pair place no restrictions on the value of the domains of propositions which may serve as restrictors of the modal operator (i.e. the D-value is unspecified), whereas the semantically encoded content of *can* and *should* includes information to that effect (indicated in the subscripts). In the former pair of verbs, the empty slot in their lexical semantics has to rely on on-line processes of pragmatic comprehension for completion; in the latter pair, a semantically specified restrictor offers a conceptual search-space, which can be further narrowed down pragmatically, if necessary. In this sense, to recall a distinction made in 1.2.1, *may* and *must* require *pragmatic saturation* of an unspecified semantics, whereas *can* and *should* may undergo *free pragmatic enrichment* of an already complete, albeit vague, semantic content.[15] Alternatively, *may* and *must* are cases of *domain selection*, while *can* and *should* are examples of *domain restriction*.

The motivation for drawing this distinction may not be immediately obvious. In the remainder of this section, I wish to demonstrate that saturation and free enrichment are the pragmatic consequences of two truly distinct and independently motivated types of semantic underdeterminacy in modals. In order to achieve this, I initially present arguments to the effect that saturation cannot be reduced to free enrichment. I then go on to show that the two processes differ, in that enrichment operates freely (i.e. optionally), whereas saturation applies in an obligatory fashion.

An alternative approach to the semantics of the English modal verbs would be to treat all modals as complete (albeit still underspecified) semantic operators. I will use *must* to illustrate. Rather than assuming that the semantic representation of the verb contains an empty slot for the type of restrictor, one might assume that it lacks such a slot altogether. On this view, the semantic entry for *must* would look like (31):

(31) ***Must***: p follows from D

According to (31), the proposition embedded under *must* follows from the set of propositions in D, regardless of the value of D (assuming that D satisfies logical consistency, i.e. includes no contradictions). This position underlies all semantic accounts of *must* which analyse the verb in terms of absolute (or logical/alethic) necessity (cf. the Modal Fallacy noted in Bradley and Swartz, 1979: 331): on this view, the various types of necessity which *must* is capable of conveying in natural language are the products of pragmatic narrowing of the broad semantic content of the verb. This position, then, essentially reduces saturation to a version of free enrichment: I will call it the *maximal restrictor* solution.

If generalised to quantificational devices in natural language, the maximal restrictor solution would imply that maximal quantificational domains form the semantic default value for a number of expressions.[16] Bach (1994a) explicitly takes this view with respect

[15] The proposed semantics for modals is not intended to be decompositional. It merely serves to illustrate the ways in which pragmatic processes contribute to the interpretation of modality in context.

[16] Alternatively, this solution could imply that the context of utterance delimits the domain of quantification in natural language without affecting the descriptive content of quantifiers. This is called the 'implicit' approach by Neale (1990), who argues against it and cites Quine (1940), Sellars (1954), Vendler (1967), Lewis (1973a), Cresswell (1973) and Grice (1981) as preferring some version of enrichment over the implicit view; cf. also Westerstähl (1985).

to another quantifier, *everyone*, when he notes that an utterance such as (32) literally (i.e. semantically) conveys (33a) and only non-literally (33b):[17]

(32) Everyone is going.

(33) a. Everyone in the universe is going.
b. Everyone in the class is going.

Maximal restrictor analyses of *must*, while favoured in logical and philosophical analyses of absolute (alethic) necessity, are too strong for purposes of accounting for the ordinary uses of the verb: on linguistic grounds alone, such interpretations are just one possibility among the range of interpretations which the verb may contextually accept. Two types of evidence argue against the maximal restrictor view. In the first place, this approach maintains that, although *must* encodes a complete concept, this concept is rarely communicated. This introduces a considerable computational complication: the hearer is obliged to access, test and, in most cases, reject the 'literal', absolute interpretation of the modal before opting for free enrichment. In this way, he invariably has to go through something which is false in order to arrive at the proposition which the speaker intends the modal utterance to convey.

More importantly, the maximal restrictor approach entails that alethic interpretations should have a facilitation effect on the comprehension of utterances containing *must*. Quite to the contrary, it seems that such interpretations have a restricted distribution in natural language data (Karttunen, 1972; Horn, 1972; Lyons, 1977; Palmer, 1990). For instance, the most natural interpretation of (34) is paraphrasable not by (35a) but by (35b):[18]

(34) The criminal must still be in the area.

(35) a. In view of everything that is logically possible, the criminal must still be in the area.
b. In view of everything that we know, the criminal must still be in the area.

---

[17] There are differences between the types of quantification in *everyone* and *must*, which are not relevant in this context.

[18] I will have more to say about the pragmatic role of the maximal restrictor in modal environments in section 2.4.3.

For similar reasons the maximal restrictor solution is problematic even if it is construed in a weaker way, i.e. as postulating pragmatic (rather than semantic) defaults for modals like *must*. In more general discussion of quantification, Roberts (1995: 691-2) seems to be advocating this view:

> It is important to note that there are examples involving all types of English operators in which the intended domain is apparently the entire relevant type of entity in the model, be it individuals, worlds, properties, or times/events. [...] Such examples suggest that these maximal domains are always familiar and frequently salient. Another way of saying this is that there is no such thing as a completely null context – the maximal domain for all types of operators is always accessible. As in anaphora resolution, where plausibility and perhaps other factors play an important role as logically accessibility, the retrieval of the intended domain depends on plausibility as well as on familiarity and salience. In examples where the relevant maximal domain is not plausible as the intended domain for an operator, if the operator has no explicit restrictive term, then felicity requires that there be some other salient and familiar domain restriction to satisfy the familiarity presupposition of the free variable R [the operator].

If the previous arguments are correct, it follows that the proposal to replace the saturation view for modals like *must* with a version of free enrichment cannot go through.

There is another way in which the distinction between saturation and free enrichment may be undermined. Suppose that we keep the proposed semantics for English modals but maintain that *must* encodes a complete (albeit vague) semantic content. This move brings saturation closer to free enrichment, in that the former process is no longer obligatory. This approach predicts that - in principle at least - there should be occasions on which *must* might express just its bare semantic content (i.e. an unspecified type of necessity). I will call this the 'rampant indeterminacy' thesis. Some authors seem to adopt this thesis. Coates (1983), among others, has argued that it is possible to understand a modalised utterance without having to decide on the kind of modality expressed.

One difficulty with this view is that pragmatic indeterminacy is exhibited by all modals, not just those of the *must/may* type. More importantly, none of the types of vagueness usually recognised for modals is particularly suitable for representing the rampant indeterminacy thesis. Consider the following example (taken from Coates, 1983: 145):

(36) The quality of the final product must be influenced by the quality of the raw material, and the methods of processing may influence its nutritional quality.

Since the two modals in (36) can receive either a root or an epistemic interpretation, the utterance is to some degree pragmatically indeterminate. There are two construals of the term 'indeterminacy' which have a bearing on the interpretation of (36). According to the first, indeterminacy arises simply because the utterance cannot be unequivocally interpreted in isolation; further contextual information, though, would make it possible to distinguish which interpretation was the one intended by the speaker. Since this is only superficially a case of indeterminacy, I will not consider it any further. On the second, more interesting construal, indeterminacy arises from the fact that both root and epistemic interpretations have (roughly) the same degree of accessibility for the hearer and are capable of causing a similar range of cognitive effects; hence, pragmatic (relevance) considerations cannot adjudicate between them even in the presence of a fuller context. The question which now arises is: could this type of indeterminacy be viewed as a case where what is communicated is the bare (unsaturated) modal meaning delivered by the semantics - which is naturally going to be perceived as indeterminate between various candidate interpretations?

The answer has to be: no. Notice that the rampant indeterminacy position makes no reference to possible pragmatic developments of the semantic content of the modals. The point of the example was precisely to come up with a case where the rampant indeterminacy thesis would make a prediction which would be unique to that thesis, and therefore critical in the comparison between a ‘rampant indeterminacy’ account and my original proposal. However, it seems that this total indeterminacy does not correspond exactly to the situation in (36): it is not so much that the utterance is indeterminate *tout court*, but rather that it is indeterminate between two alternative (but specific) ways of narrowing down the semantic content of the two modal verbs. In fact, I doubt that there can be any case where total indeterminacy exists in modal utterances (for further arguments and evidence for a moderate type of indeterminacy, see 2.3). Consequently, it is not the case that the rampant indeterminacy solution is in a position to make predictions which the saturation view for modals cannot; moreover, it seems to make the wrong predictions by allowing for interpretations which never arise in a natural way.

To conclude: rather than symmetrifying the picture of the semantics of modals by reducing saturation to free enrichment, it seems that both options have to be maintained to deal with individual cases. In the next section, I demonstrate how different semantic inputs are pragmatically processed to yield the variety of interpretations of modal verbs.

2.3 THE PRAGMATICS OF ROOT MODALITY

2.3.1 Derivation of Root Interpretations

Consider (37):

(37) Computer-aided instruction can co-occur with more traditional methods of teaching.

The logical form of the utterance is given in (37'):

(37') $_p$[Computer-aided instruction co-occurs with more traditional methods of teaching] is compatible with $D_{factual.}$

According to the proposed semantics for *can*, the verb lexically specifies as its restrictor the domain of factual propositions. This semantics captures the widely shared intuition that *can* encodes potentiality (Walton, 1988; Bolinger, 1989; Klinge, 1993): a state of affairs is characterised as potential when it is compatible with the states of affairs in the actual world, and hence may itself be actualised at some point in the future. The restrictor in *can* may be further contextually narrowed so as to pick out a sub-domain of factual assumptions. This is illustrated by the possible paraphrases of (37) given below:

(38) a. In view of the way schools are run these days,
b. In view of the technical equipment available for education purposes,
c. In view of the teachers' encouraging stance,
computer-aided instruction can co-occur with more traditional methods of teaching.

Each of the above paraphrases captures a different concept of potentiality, which may be constructed 'on the fly' during the comprehension of (37). These *ad hoc* concepts correspond to restricted interpretations of modality constructed from different sets of factual premises. Intuitively, a crucial component in the construction of these interpretations is deciding on the sort of factors which determine how school education is run. The variability of these factors is responsible for the variability in the modal readings in (37).

The exact content and recovery of the modal restrictor pose the most interesting problems for the pragmatics of modality. Generally, the intended (sub)domain for the comprehension of a modal has to contribute to an interpretation of the utterance which is accessible enough for the hearer, and capable of achieving adequate cognitive effects in a way compatible with the speaker's abilities and preferences (i.e. an optimally relevant interpretation); furthermore, the resulting interpretation should be one that the speaker could reasonably have intended to be optimally relevant for the addressee. In recovering the restrictor for a modal expression, the hearer, therefore, typically makes use of assumptions which are easily accessible from the encyclopedic entries of the concepts in the complement proposition and other assumptions which are contextually available. Moreover, since communication raises specific expectations of relevance, the hearer aims at reconstructing the type of background propositions the speaker has in mind. Suppose that, in (37), the interlocutors have been discussing the changes in society brought about by technological advances. In this case, the intended domain for the comprehension of modality involves assumptions about technological progress: what the utterance broadly conveys is that facts about technology have opened up new possibilities in education. The intuition that the intended domain for modality has to be easily recoverable from the context has been known as the 'familiarity presupposition' for domain selection – see Roberts (1995). I will return to this issue later in this section in discussing some preferred interpretations for *can* and *may*.

Notice that each of the paraphrases in (38) is a shorthand for the full set of assumptions included in the intended domain for modality. These assumptions form a web of interconnected mental representations concerning education, current teaching practices, etc., and ultimately depend on law-like generalisations about how facts are causally related. An attempt to spell out all of these assumptions would soon be frustrated.[19] It is, therefore, important to emphasise that the comprehension of modality does not involve recovery of the full set of propositions which form the modal domain - just a broad *description* of the domain, or a specification of the *type* of propositions the domain includes. *A fortiori*, it follows that the recovered modal restrictor does not feed into the proposition expressed by the modal utterance in the form of a set of background propositions. Rather, the specification of the intended domain takes the form of a general description, which is normally pragmatically inserted in the proposition expressed and affects the truth conditions of the modal utterance.

---

[19] There are certain similarities between Searle's (1983) notion of Background and the content of domains, but Searle insists that the Background is non-representational in nature.

There are several arguments which show that at least such a broad description of the intended modal domain (i.e. of the intended type of modality) is necessary for the comprehension of a modal utterance. A first argument is offered by the presence of jokes and misunderstandings. Consider the following example (cited in Kratzer, 1981a: 53). A philosopher once gave a lecture in which he mentioned that one of the defining properties of a cup is that one can pour things like coffee in it. A student objected that – if this were true – a cup which has coffee in it already would not be a cup anymore.

More generally, misunderstandings are a good indication that there has been a mistake in the recovery of the modal restrictor. Assume that a zookeeper addresses (39) to his new assistant:

(39) The monkey can climb to the top of the tree.

Later on in the day, the monkey is missing. The zookeeper is angry at his assistant: he says he warned him that the monkey was capable of climbing to the top of the tree and so could escape. The assistant replies that he interpreted the utterance to be about what the monkey was allowed to do, and was therefore not worried when the monkey behaved in just this way. The misunderstanding is due to a domain mismatch: the zookeeper had in mind potentiality in terms of the monkey's physical abilities, while the assistant had in mind potentiality in terms of the zoo's regulations.

A further piece of evidence which shows that interlocutors are sensitive to subtle aspects of the modal restrictors is that speakers and hearers frequently shift and modify modal domains during a single conversational exchange. Imagine that Alice and her lawyer have been discussing the prospect of Alice's having a divorce. Alice utters (40) and her lawyer replies as in (41):

(40) I can't leave my husband penniless.

(41) Of course you can - the law allows you to.

The modal domain in (40) includes assumptions about Alice's feelings and moral strength, whereas in (41) it includes assumptions about legal regulations. This example is a clear demonstration of how the restrictor affects the truth-conditional content of a modal utterance. The shift in the restrictor is an instance of accommodation (Lewis, 1979), where there is a change in the boundary of the relevant domains for modality. The capacity to perform appropriate shifts in the modal restrictors is on a par with the more general ability to pragmatically infer the intended domain for a modal operator by monitoring cost-effect expectations which the speaker has built into the modal utterance.

The above arguments go against the view (discussed in the previous section) that modal expressions standardly exhibit rampant indeterminacy: as it turns out, in a large class of cases, modal expressions are understood in a fairly specific and restricted way. Nevertheless, it is important to recognise that there is considerable variation in the degree of specificity of the modal restrictor. On certain occasions, modals communicate a broad notion of possibility or necessity, while on others, more fine-grained modal concepts are called for. Consider the example in (42) with the possible rejoinders in (43):

(42) Do you want quick and accurate medical advice?

(43) a. Now you can have it.
b. Now you can have it free on our web site.

Suppose that the utterances are part of an advertisement. What *can* communicates in (43a) presumably coincides with the encoded content of the verb: it is compatible with a set of factual propositions that the public has quick and accurate medical advice. In (43b), the modal conveys a more restricted type of potentiality, namely potentiality in terms of a company's setup and electronic services. The present account offers a natural and original way of predicting when each of the rejoinders will be felicitously produced: the extent of specificity of the modal restrictor will depend on expectations of cognitive effects. In (42a), it is hard to imagine that a broad potentiality reading will contribute to an interpretation of the utterance which is capable of achieving adequate cognitive effects. The hearer is left wondering how he can access the medical advice advertised, and, therefore, misses part of the point of (43a). In (43b), the PP *on our web site* provides the restricting domain (the enabling conditions) for the comprehension of the utterance. By providing the grounds for the modal statement, (43b) fulfils the hearer's expectations of relevance: it brings about a range of cognitive effects (e.g. contextual implications about how diagnosis or treatment are transformed through the Internet) at no unwarranted processing effort. It follows that, in the above context, (43b) will be more felicitous than (43a).[20]

[20] A further consequence of this account is that, in contexts in which two candidate restrictors are equally accessible and have no significant difference in terms of cognitive effects, some moderate indeterminacy in the modal will be tolerated. In those cases, the interpretation of the utterance oscillates between the two domains, without having to be resolved in favour of one or the other. Coates' (1983) cases of 'merger' belong here - I repeat an earlier example:

Similar pragmatic considerations are responsible for certain constraints on domain-shifting. Consider the examples in (44)-(45):

(44) Alice: We cannot go on living together.

(45) Her husband: We can – for the sake of the children.

Once (45) is uttered, Alice cannot keep maintaining (44) without further justification: a new possibility has been raised and cannot be ignored. Put differently, what Alice said in (44) was true with respect to the modal domain she had in mind, but is no longer true with respect to the modal domain introduced in (45) (involving the children's well-being), and so her husband seems to have the last word. This observation was first made by David Lewis who noticed that, once the boundary of modality (i.e. the modal domain) shifts outward, it remains shifted. He notes (Lewis, 1979/1991: 425):

> Again this is because the rule of accommodation is not fully reversible. For some reason, I know not what, the boundary readily shifts outward if what is said requires it, but does not so readily shift inward if what is said requires that. Because of this asymmetry, we may think that what is true with respect to the outward-shifted boundary must be somehow more true than what is true with respect to the original boundary. I see no reason to respect this impression. Let us hope, by all means, that the advance toward truth is irreversible. That is no reason to think that just any change that resists reversal is an advance toward truth.

The reason why a domain shift, once performed, cannot be easily cancelled, is due not to truthfulness but to appropriateness considerations. Successful communication requires co-ordination between the interlocutors as far as the recovery of the modal restrictor is concerned. In the case of outward domain shifting, as in (45), a change is performed in the interlocutors' mutual cognitive environment: the intended grounds for a modal utterance are enlarged through the introduction of further assumptions. Unless these further assumptions are explicitly removed, they will remain in the mutually manifest context in the next stages of the exchange and will affect the grounds for assessing whether it is possible for the interlocutors' to go on living together.

(i) The quality of the final product must be influenced by the quality of the raw material, and the methods of processing may influence its nutritional quality.

So far I have given an illustration of how the proposed model for modality works, drawing examples mostly from the 'potentiality' reading of *can*. I now want to extend the analysis to the rest of the possible interpretations of this verb (and to other modals).

The 'ability' reading of *can* has captured the interest of many commentators. A number of them have argued that this is a non-modal use, which corresponds to the basic meaning of the verb (see, e.g., Foolen, 1992). I want to take the view that the 'ability' interpretations of *can* are as much the result of enrichment of the semantic content of the verb as the various 'potentiality' interpretations of (37). Consider an earlier pair of examples, repeated below as (46a-b):

(46) a. Mary can speak German.
b. Mary can speak German at the meeting, because everybody will understand.

In its most natural interpretation, the utterance in (46a) will be taken to communicate something about Mary: the hearer will retrieve (or construct on-line) the encyclopedic entry for Mary and process the embedded proposition $_{p}$[Mary speaks German] against assumptions that have become available through the activation of this encyclopedic entry. In other words, the state of affairs described in p is taken to involve a property of Mary's which is compatible with her other features. This way of narrowing down the factual domain of propositions in (46a) differs from (46b). In order to compute whether p is compatible with the relevant set of propositions in the factual domain, we now need to take into account a broader sub-domain of factual propositions apart from those involving Mary's properties - for instance, assumptions concerning the situation in meetings, the other participants, and so on. Therefore, we cannot get a pure 'ability' reading for (46b) as we can for (46a).

I suggest, then, that ability interpretations for *can* arise whenever the sub-domain of factual assumptions which is taken to be compatible with the embedded proposition belongs to the 'file' for an individual or object (which also normally appears as the sentential subject). This analysis predicts that, whenever both individual-specific and broader factual considerations bear on the comprehension of *can*, the traditional ability/potentiality distinction will collapse - this is exactly what happens in (46b). By resisting proposals to place 'ability' concepts within the semantics of *can*, this account also manages to avoid a host of problems which these proposals face. For instance, inanimate subjects and passive sentences, which provide counterexamples to an ability-based semantics for *can*, are naturally predicted to favour a root potentiality interpretation on my account: there is no individual or object to furnish a fact-supporting conceptual (:encyclopedic) entry, with which the new attribute would be judged compatible:

(47) a. Bad weather can ruin the crops.
b. The total sum can be divided into two equal parts.

Moreover, the present account offers a satisfactory explanation of the relation between *can* and *be able to*; the latter is taken to encode ability, as shown in the following contrasts:

(48) a. John can/?is able to swim, if he likes.
b. We can/?are able to offer you a discount, if you wish.

Unlike what the conditionals convey, inherent ability cannot be subject to an individual's wishes - John's in (48a) or the addressee's in (48b). This incongruity is the reason for the unacceptability of the utterances containing *be able to*. Utterances containing *can*, on the other hand, do not give rise to similar problems since ability is not present in the semantic content of the verb.

Can is used 'deontically' to communicate permission in cases in which it is mutually manifest that the proposition p expressed by the embedded clause describes a state of affairs that is desirable from the hearer's point of view, it is within the hearer's power to bring about this state of affairs, and the speaker has authority over the hearer:

(49) You can go home.

Finally, *can* may be used to convey a suggestion or an offer. This type of interpretation requires the presence of the following collection of contextual assumptions: the embedded proposition p represents a state of affairs which is manifestly desirable to the hearer from his own point of view, as well as beneficial to him; the speaker has the responsibility for bringing about the state of affairs described in p; and the speaker manifestly lacks any obligation to bring about this state of affairs:[21]

(50) a. We can meet one day after work.
b. I can give you a lift.

---

[21] On alleged epistemic uses of *can*, see 2.4.2. I will not deal here with uses of *can* (or other modals) in indirect requests of the sort *Can you pass the salt?* – see Papafragou (2000) for discussion.

I now turn to *may*. Imagine that (51) is uttered by a bank employee to a young customer:

(51) Our branch may convert your account into a student account; you just need to supply us with proof of student status.

The utterance has the logical form in (51'):

(51') $_p$[Our branch converts your account into a student account] is compatible with D.

This logical form requires some fleshing out before it can be a truth-evaluable representation. Among other things, it remains to be specified what is the value of the domain D. Suppose that (51) is an answer to the customer's query about a change in the status of his account. In such a context, the customer has made mutually manifest that he is interested in changing the type of account he has, i.e. that he considers such a change desirable from his own point of view. As a result, it already forms part of the two interlocutors' mutual cognitive environment that the change is compatible with the speaker's preferences. The only accessible enough domain of assumptions for which it would be relevant to know whether it is compatible with p involves the bank regulations. As a result, (51) is interpreted as informing the customer that the bank regulations permit the branch to convert his account into a student account.[22]

Similar considerations apply to examples (52) through (54):

(52) I'll tell you about your journey, so that you may make arrangements.

(53) To make this dish, any sort of pasta may be used.

(54) During the seminar, you may interrupt as often as is needed.

In (52), the domain of propositions with which the embedded proposition is compatible is a sub-domain of factual propositions describing the hearer's abilities and general circumstances concerning his journey. In (53), this domain is provided by the content of a recipe (a regulatory domain). In (54), the situation is a little more complex. Suppose that the utterance is produced by the person teaching the seminar to the class. Then it is mutually manifest to the interlocutors that students are in a position to interrupt as long

[22] For purposes of the present discussion I disregard the fact that the example may also receive an epistemic reading.

as the teacher does not object. It follows that the only domain of propositions that could relevantly contribute to an interpretation of the modal verb in (54) is that concerning the teacher's own preferences. Given two further, mutually manifest contextual assumptions, namely that interrupting the seminar as often as is needed is desirable from the students' point of view, and that the teacher has some authority as far as the seminar is concerned, (54) will be felt to convey a 'permission' interpretation.[23]

One might observe a common pattern in the way *can* and *may* contribute to the relevance of utterances in which they occur. In many cases, the context includes an assumption to the effect that the embedded proposition p is incompatible with the set of propositions in a certain domain D (or assumed to be so). What *can* and *may* do, then, is to contradict and eliminate this contextual assumption, by conveying that p *is* compatible with the propositions in D - or, to put it differently, that the negation of p (~p) does not follow from the set of propositions in D. This explains how utterances containing *can* or *may* are capable of achieving any cognitive effects at all, given that their communicated content (a compatibility relation among propositions) makes for a very weak statement.

Notice that, on a Gricean analysis, utterances containing *can/may* come out as informationally impoverished, on a par with other constructions such as negative utterances (Karttunen, 1972; for a neo-Gricean account of negation, see Horn, 1989). The parallel is to be expected: if we consider compatibility as an implicitly negative term,[24] the cases of *can/may* fall squarely within the province of negative operators. As is generally acknowledged, utterances containing a negation marker can be felicitously used only in special contexts: in discussing Pierre's news, for instance, it would be awkward to assert (55) in the absence of a contextual assumption to the effect that (it is expected that), after his divorce, Pierre will visit his daughter whenever he likes. Similarly for *can/may*, it would be awkward to assert (56) in the absence of a contextual assumption to the effect that (it is expected that) his divorce will prevent Pierre from visiting his daughter whenever he likes:

(55) After his divorce, Pierre will not visit his daughter whenever he likes.

---

[23] I consider *may* in the following (rare) structures a separate lexical item (examples from Quirk *et al.*, 1985):

(i) May the best man win!
(ii) May he never set foot in this house again!
(iii) May God bless you!

[24] The term 'implicit negative' belongs to Fodor *et al.* (1975) and refers to terms whose underlying semantic representation includes a negative term. Their examples (*deny, doubt*) differ from *can/may* in that their syntactic behaviour parallels that of explicit negatives (e.g. they trigger negative polarity items).

(56) After his divorce, Pierre can/may visit his daughter whenever he likes.

What this fact about negation suggests, according to a number of psychologists and linguists (Wason, 1965; Greene and Wason, 1970; Givón, 1978; Horn, 1989), is that, on their commonest and most natural interpretations, negative statements are understood as denials. In this sense, negatives (and, by extension, *can/may*-utterances) in some pragmatic sense presuppose the existence in the immediately accessible context of their affirmative counterparts.

In relevance-theoretic terms, the weak informativity of utterances of modal possibility can be redefined as non-productivity of contextual implications. In (56), the explicatures of the utterances are unlikely to interact with any existing contextual assumptions to yield contextual implications. Given that they are also unlikely to strengthen any existing contextual assumptions, it follows that the only way open to them for achieving relevance is by contradicting and eliminating an assumption previously held in context. This typical path in the interpretation of *can* and root *may* is precisely what my analysis in the previous paragraphs has suggested.

It seems, then, that an important class of possibility statements are intended to reject a previous assumption, according to which the proposition embedded under the modal operator describes an impossible state of affairs. What the possibility utterance does is encourage the hearer to concentrate on the type of evidence which was previously assumed to exclude the proposition expressed by the embedded clause, and invite him to reconsider it. While unique to relevance theory, this proposal is capable of capturing intuitions discussed in other accounts. For instance, it makes available a pragmatic explanation for Sweetser's (1990) idea that *may* encodes the existence of a potential but absent barrier. What the 'barrier' idea taps is that, although the proposition p which is embedded under *may* was (or could be) previously contextually expected to be incompatible with the assumptions in the appropriate propositional domain, this is not in fact the case.

There is a second way in which *can/may* contribute to relevance. Both verbs may be used to raise a possibility which was previously absent from the context. In cases in which the embedded proposition lacks any status (i.e. is considered neither possible nor impossible), a modal utterance with *can/may* assigns to that proposition a potentiality/possibility status. Assume that (56) is part of the court ruling about Pierre's divorce. Before the ruling, neither p ('Pierre will visit his daughter whenever he likes') nor ~p ('Pierre will not visit his daughter whenever he likes') are part of the communicators' mutual cognitive environment; moreover, it is not known whether any of the two propositions is compatible with Pierre's rights and obligations. (56) establishes that it is compatible with Pierre's rights and obligations that he visits his daughter whenever he likes. This new piece of information can interact with a contextual assumption such as (57a) to yield a contextual implication such as (57b):

(57) a. If, after the divorce, Pierre can/may visit his daughter whenever he likes, his ex-wife will be very upset.
b. Pierre's ex-wife will be very upset.

In a similar vein, *can* in offers and suggestions is normally interpreted as introducing possibilities where there were previously none:

(58) I can carry your suitcase.

I conclude the comparison of *can* and *may* by surveying their permission uses. A number of authors have observed that, in such uses, the two modals are occasionally but not always interchangeable. The differences between the verbs are brought out in (59):

(59) a. You can smoke in this room.
b. You may smoke in this room.

Suppose that the following contextual assumptions are mutually manifest to the speaker (Monica) and hearer (Patrick): (a) the state of affairs in which Patrick smokes in the room is considered desirable from his own point of view; (b) it is within Patrick's power to bring about this state of affairs; (c) Monica has some sort of authority over Patrick. By using (59a), Monica informs Patrick that there is nothing in the factual domain which blocks his smoking in the room. In other words, circumstances are such that he is allowed to smoke. By contrast, (59b) can be used to communicate that Patrick's smoking in the room is compatible with a variety of assumptions, including e.g. the speaker's preferences. This is the reason for generally considering *may* a more polite form for giving permission than *can* (since it conveys, in a sense, greater speaker involvement).

It is not the case that *may* is always felt to be polite. Imagine that (60) is uttered by a general after interrupting a soldier who is talking: the soldier will feel under greater obligation to go than if he had simply received permission:[25]

---

[25] Such examples have been treated in the literature as cases of 'pragmatic strengthening' which form part of 'context-induced reinterpretation' within a grammaticalisation chain (Heine, 1995).

(60) You may go home, Jones.

Here, the context lacks the assumption that Jones considers going home as desirable from his point of view (I will call this assumption z); actually, the fact that Jones was in the middle of talking rather makes mutually manifest the opposite assumption (i.e. that he does not wish to leave immediately). However, it is difficult to see how the general could have intended (60) to be optimally relevant to Jones, unless the general believed that z is true. So Jones has to supply z as a background assumption and attribute it to the speaker (even though he knows z is false). The interpretation would proceed along much the same lines as (59a) above, except that Jones knows that the general knows that z is false: the general makes as if z were a mutually manifest contextual assumption. The intuition that, under these circumstances, the utterance is rude, results from the fact that the general ignores the hearer's preferences, although these are contextually salient. A similar analysis can be given for the following examples with *can*:

(61) a. You can forget about your Christmas present - you haven't been a good boy.
b. You can start looking for a new job.

I now want to move on to the pragmatics of *must*. Imagine that Mary finds herself in a very cold room and utters (62):

(62) I must sneeze.

The logical form of the utterance is provided in (62'):

(62') $_{p}$[Mary sneezes] is entailed by $D_{unspecified}$.

What the hearer is expected to retrieve is that Mary's sneezing is a necessary outcome given her physical condition and the circumstances in the room. This entails that the unspecified domain D in (62') has to be pragmatically narrowed down to a sub-set of factual propositions. Other types of contextual enrichment of the semantic content of *must* are given in (63)-(65):

(63) In opening a game of chess, the players must move a pawn.

(64) The President must formally approve the new Government before it can undertake its duties.

(65) The accused must remain silent throughout the trial.

These three utterances require regulatory domains of different types. (63) expresses a necessity with respect to the rules of chess, (64) a necessity with respect to the Constitution and (65) a necessity with respect to judicial rules.

Must admits a variety of propositional domains as restrictors. (66) is a slightly more complex example:

(66) I must lose weight.

Assume that the speaker, Amy, wants to become attractive and realises that the only way to achieve this is by losing weight. It follows that losing weight is necessary in view of Amy's desire to become attractive (and the factual assumption that, unless she loses weight, she will not do so). (66), then, is the product of (one version of) what philosophers have called *practical syllogism*, i.e. a kind of syllogism which is not concerned with belief-formation and persuasion but with intention-formation and action (von Wright, 1963; cf. Anscombe, 1957; Harman, 1976; Kratzer, 1981a, 1991). Practical syllogism has the following general form:

(67) a. I want to attain y.
b. Unless z is done, y will not be attained.
Therefore, z must be done.

Applied to (66), (67) yields (68) or, more formally, (69):

(68) a. Amy wants to become attractive.
b. Unless she loses weight, Amy will not become attractive.
Therefore, Amy must lose weight.

(69) a. $_{q}$[Amy becomes attractive] $\in D_{des}$,
where D_{des} = what is desirable for Amy
b. $\{_{p}$[Amy loses weight] $\in D_{factual}\} \vee \{\sim_{q}$[Amy becomes attractive] $\in D_{factual}\}$
Therefore, p is entailed by D_{plan},

where $D_{plan} = D_{des} \cup D_{factual.}$[26]

D_{plan} can be paraphrased as the domain consisting of Amy's plans. What (69) captures is that, for Amy, losing weight is a necessary means to a desirable end. This type of necessity is commonly conveyed by *must*. This account fares better, I think, than previous approaches to deontic *must*. For instance, Groefsema (1995) has proposed that the proposition expressed by utterances such as (66) becomes enriched to the point that it represents a desirable state of affairs. (66), on her analysis, comes out as having the following enriched propositional form (assuming my semantics for the modal):

(70) Amy desires that $_k\{_p$[Amy loses weight] is entailed by $D_{unspecified}\}$.

As I have already pointed out, however, (70) does not correspond to an intelligible desire. Moreover, what Amy desires is effectively not the state of affairs described by k but a different one - that she becomes attractive - for which losing weight is a prerequisite (see section 2.1.3.2). On my account, the proposition which the speaker considers desirable will not be directly detachable from the surface structure of the utterance in (66).

There is only a short distance from the type of interpretation in (66) to the purely deontic (i.e. obligation-imposing) uses of *must* in (71):

(71) a. You must write 100 times 'I will never yawn in class again'.
b. You must love your fellow humans.

The deontic interpretations of (71) arise in case (i) the modal restrictor involves a set consisting of the speaker's desires and factual assumptions (or, alternatively, a set of regulatory propositions which the speaker is entitled to enforce); (ii) the speaker has authority over the hearer; (iii) the hearer is in a position to bring about the state of affairs described in the embedded proposition. An interesting variant of the 'obligation' interpretation arises in cases of quasi-imperative suggestions/offers (cf. section 2.1):

(72) a. We must go for a drink one day.
b. I absolutely must walk home with you.
c. You must come and visit us sometime.

[26] Talking of union of sets is a simplification; see section 2.3.2 for discussion.

Consider in particular (72a). The utterance conveys that it is entailed by (the appropriate sub-set of) the speaker's desires that she and the addressee go for a drink one day. Since the addressee most probably has to show sensitivity to her interlocutor's desires due to the social rules governing their relations, the utterance will be interpreted as an urgent form of suggestion (for the contextual background against which 'suggestion' readings typically arise, see the discussion of (50) above).

I come, finally, to *should*. The verb is often felt to convey obligation of a weaker type than *must* in examples like the following:

(73) You should clean the place once in a while.

(73') $_{p}$[You clean the place once in a while] is entailed by $D_{normative}$.

According to its semantically specified restrictor, *should* expresses a necessity relative to existing stereotypes, or norms. The comprehension of *should* relies quite heavily on the sort of structured knowledge humans typically possess about the normal course of events, which has been referred to by various writers in cognitive psychology as 'scripts', 'frames', 'scenarios', and so on (see 1.2.1). To the extent that norms, apart from being statistical generalisations, acquire regulatory status as well, root interpretations of *should* become indistinguishable from certain interpretations of *must*. Still, since what is expected/normal can be quite different from what is commanded, *should* is often seen as communicating a less urgent kind of necessity than *must*.

From that point of view, the verb bears certain similarities to *ought to*, which, I suggest, semantically specifies as its modal restrictor the domain of ideal/morally recommended states of affairs:

(74) One ought to respect one's country.

(74') $_{p}$[One respects one's country] is entailed by D_{ideal}.

(75)-(76) illustrate the differences between the types of necessity communicated by *must*, *should*, and *ought to*:

(75) In this game, you must/?should/??ought to carry an egg in a spoon and be careful not to drop it.

(76) Chief scout to the younger boys:
a. You must be back by midnight, ?although it's fine by me if you aren't.
b. You should be back by midnight, (?) although it's fine by me if you aren't.
c. You ought to be back by midnight, although it's fine by me if you aren't.

In (75), since *must* lacks any semantic conditions on the restrictor, it is open to an interpretation where the rules of the game function as the restricting domain. The utterance with *should* is slightly worse: although the rules of the game can do duty as a normative domain, this particular game is far removed from what is a typical game activity (compare: *In this game, you should try to reach the end faster than your opponent*). Finally, *ought to* in (75) is distinctly odd, since the domain semantically provided for the interpretation of the verb clashes with the pragmatic requirements on the comprehension of the utterance.

In (76), the differences in acceptability also bear on the selection of the restrictor: (76a), on its most accessible interpretation, is construed as imposing an obligation on the younger scouts (given a number of contextual assumptions concerning authority and social relations in the group); the continuation of the utterance therefore becomes unacceptable. (76b), although it can receive a similar interpretation, is also open to another reading: the chief scout simply reports what is entailed by the norms/expectations concerning a scout's behaviour, but is manifestly not in agreement with those norms, so that he will not use his authority to enforce them. Finally, (76c) is a mere statement concerning what ideals or duty entail, which can differ substantially from what the speaker is prepared to accept in practice.

It is worth noting here that, in accordance with a number of researchers in the field (see Warner, 1993), I assume that *should* is not synchronically perceived as the past tense form of *shall* but has entered the modal system as a separate, individual item. Still, there are a number of uses of the verb which are considered to retain overtones from its original meaning, or to be generally unrelated to the rest of its root interpretations (Ehrman, 1966; Coates, 1983; Palmer, 1990). For instance, *should* is used in ways similar to hypothetical *would* as in (77a), as a hypothetical marker with inverted word-order as in (77b), or with a seemingly null contribution to meaning (i.e. as a place-holder) in the quasi-subjunctive construction in (77c):

(77) a. I should be grateful if you could bear my case in mind.
b. Should you require any further assistance, please feel free to contact us.
c. I do not desire that I should be left alone in this task.

These examples involve hypothetical environments with varying degrees of complexity. Some might be dealt with rather straightforwardly within the existing framework; the first one, for instance, requires us to add an extra condition to the 'expectation' restrictor of *should*, namely the antecedent of the conditional. The other examples represent more idiosyncratic aspects of the English modal system, which probably have to be learnt individually. This is especially true for (77b), where the use of *should* correlates with a distinct syntactic configuration peculiar to auxiliaries (which is parasitic on the semantics of the conditional).

This raises a more general issue with respect to the scope of the proposed account of the root interpretive class. Although I have focused on the central readings of modals, there are some peripheral cases that resist being subsumed under semantic generalisations. To repeat a point made in section 1.1.3, a system as complex as that of English modal verbs includes exceptions and individually memorised uses of specific modals. Such uses have to be stipulated on any semantic-pragmatic account of the modals and therefore do not constitute a particular difficulty for the unitary semantic thesis I have been pursuing.

2.3.2 Issues Related to Modal Restrictors

In the previous pages, I have argued that modals are context-dependent expressions, in that they depend on inferential pragmatic processes to complement the information they semantically encode. Depending on the type of inferential manipulation they permit, modal verbs are understood as communicating different kinds of modal meanings. My purpose in this section is to raise and address a potential problem for the proposed account.

One might object that the present formulation of domain-construction does not adequately characterise the formation of modal restrictors. The objection could be phrased as follows:

> Previous accounts of modality such as Kratzer (1981a, 1991) deal with the problem of domain-construction for modals by adopting two types of modal domain: a modal base and an ordering source. The interaction between these two conversational backgrounds produces a variety of modal interpretations. One of the reasons for postulating ordering sources is the fact that our reasoning about alternative realities seems to be limited by certain standards of what is ideal (or

normal). Another reason is supplied by cases of practical inference such as the following:

(78) I must work harder.

(78) intuitively involves a necessity in terms of both the realistic circumstances and the speaker's desires. Without an ordering source, Kratzer would have to deal with such examples in terms of a union of two types of modal base, one including circumstances and another including the speaker's desires. Such a union, however, would result in an inconsistent set of propositions, since wishes often clash with actual circumstances. As Kratzer points out, the postulation of ordering sources gives a way of avoiding the problem of inconsistency: (78) is handled in terms of a circumstantial modal base, whose contents are ordered with respect to closeness with what the speaker desires (a buletic ordering source).

Compared to Kratzer's account, the present account of modality is 'barer'. It recognises only one kind of modal restrictor (modal domains) which corresponds roughly to Kratzer's modal base. The question arises whether additional machinery is available (or needed) in order to deal with 'ordering source' information. Significantly, the proposed account seems to be relying on combinations of modal domains at a number of points, including cases of practical inference (cf. the discussion of example (66)). Moreover, in direct contrast to Kratzer, it is assumed that combination of modal domains can be handled in terms of union of two different domains.

In responding to this objection, I wish to argue that ordering source information is supplied by cognitive considerations of economy and plausibility and, therefore, has no proper place in a semantic account of modality. I take the facts in need of explanation to be the following: in order to deal with examples of practical inference and other cases of domain combination, hearers generally follow the strategy of constructing alternative domains adhering as closely as they can to available representations in the factual domain. That is, hearers use as a starting point a stock of non-factual assumptions (e.g. desires) and typically use pragmatic inferencing to fill out the new domain (mainly) out of factual assumptions.

Earlier on, in criticising the Lewis-Kratzer notion of similarity among worlds, I suggested that the similarity of representations of alternative realities to the representation of actual/normal states of affairs can be explained on cognitive grounds. I now want to expand a little on reasons why domains other than the factual one rely heavily for their structure on factual information (see also Papafragou, 1996a). The reasons, I suggest, relate to two-fold considerations of relevance: amount of (expected) cognitive effects as well as economy of processing effort. On the effects side, given the strong cognitive gains from handling factual assumptions, a rational being would have no reason to entertain alternative possibilities if they had absolutely nothing to do with the way the world actually is. In most cases, in order to achieve relevance, representations of a given domain should be descriptions of a state of affairs not simply possible but also potential, that is, close enough to the actual one to be able to be, become or have been actual itself.

On the effort side, it makes sense to keep representations of alternative states of affairs similar to our representation of actuality. There are obviously quantitative and qualitative aspects to similarity. On the one hand, the individual representations have to share a number of propositions; on the other, the sort of proposition they share has to be important, or non-trivial. To be sure, thoughts about alternative states of affairs may depart quite radically from our mental representations of reality. However, every departure from the current construal of the world demands considerable cognitive effort, and therefore has to be offset by adequate cognitive effects. What these effects consist of is a different picture of what the *actual* world could be (become/have been), if a set of conditions (e.g. moral rules, obligations, etc.) had been met. Therefore, the most economical way of setting up representations of alternative realities is to introduce a set of separate assumptions (ideal, stereotype-based, or other) into a body of assumptions about the real world which is held maximally constant. Moreover, apart from the cognitive side, there is a communicative side to this issue. When inviting the hearer to entertain a representation of an alternative state of affairs, and see what follows from it, the speaker should be careful to convey a representation which the hearer can reconstruct from the evidence he possesses, that is, from his own representation of actuality. Since the actual world is (by definition) the only world which is realised, it is going to be the one which furnishes a vast range of assumptions mutually manifest to the interlocutors. It would be communicatively infelicitous (Grice would say 'un-cooperative') to introduce alternative states of affairs which depart from these mutually manifest assumptions for no good reason and in unpredictable ways.

The problem of deciding how much of the structure of the actual world to preserve when constructing alternative possibilities has been much discussed in philosophy and linguistics. As Stalnaker puts it, ' some opinions acquire a healthy immunity to contrary evidence and become the core of our conceptual system, while others remain near the surface, vulnerable to slight shifts in the phenomena' (Stalnaker, 1970: 126). The former correspond to assumptions we are unwilling to part with when constructing alternative states of affairs - for instance, laws of nature, or particularly compelling empirical facts. The latter correspond to contingent details, which, if they were otherwise, would not seriously influence the picture we have about the world as it is. I will not pursue the issue any further but merely take for granted that some of our ideas about the world are cognitively more central, and therefore unlikely (because of processing cost) to be revised or abandoned when we set up plausible alternatives to actuality.[27]

I have assumed that the construction of alternative domains proceeds essentially as a union between sets of factual and non-factual assumptions. Most probably, the interaction of the two types of sets will bring about modifications, since assumptions standardly come in smaller networks of locally dependent assumptions. Take a situation in which I have two sisters instead of one. This situation will not differ from the actual one only in terms of a single state of affairs described by the proposition p 'I have two sisters': it will also differ in terms of a range of propositions which would also be true once p was true (such as 'My family has five members', or 'There are four females in my family'). This is because whatever aspect of the world makes p true also makes the rest of these propositions true. This relation between propositions (called 'lumping' by Kratzer, 1989) is obviously very important in specifying the exact content of alternative domains. However, when it comes to the psychological representation of possibilities, not all propositions lumped by a novel proposition will be computed and added to the representation of alternative realities (i.e. to non-factual domains). Considerations of processing effort will ensure that only aspects which may yield cognitive gains will be explicitly represented and processed. These cognitive gains do not include, for instance, the trivial implications of p I mentioned. Rather they include implications which are capable of entering into further inferential processes and yield cognitive effects (e.g. 'I get less attention from my parents', 'I share my room with two people', and so on). To the extent that dropping, adding or modifying assumptions about the actual world does not have any immediate relevance, we can safely suppose that it is left undone. As I have already hinted, when constructing a non-factual domain, we manipulate that part of it which is cognitively rich and assume the (unspecified) rest to resemble the factual domain, even though we recognise upon reflection that this is not strictly speaking correct.

On this view, Kratzer's ordering source on modal bases - together with the requirement to respect normalcy standards in constructing alternative worlds - fall out from more

[27] For more detailed discussion of the philosophical issues involved here, see Stalnaker (1968, 1970, 1980), Goodman (1955), Lewis (1973a).

general cognitive considerations. Furthermore, this account escapes the risk of creating logically inconsistent mixed domains, e.g. through arbitrarily carrying over factual assumptions into desirability domains (since what we consider desirable may clash with the facts). Cognitive economy prevents a fuller conceptualisation of an alternative world than strictly necessary. Still, to the extent that we need to represent an alternative state of affairs, we adjust the content of the new domain so that only the most accessible and better evidenced factual assumptions will be retained and combined with desires, obligations, ideals, and so on during the construction of non-factual domains.

2.4 The Pragmatics of Epistemic Modality

2.4.1 The Metarepresentation Hypothesis

Throughout the previous discussion I have assumed that modality operators encode a logical relation between a proposition and a set of other, contextually specified propositions. I now want to claim that what sets epistemic modality apart from the kinds of modality examined so far is that the proposition embedded under the modal here is a metarepresentation.

Let me start with some preliminary definitions, which I have already alluded to in section 2.2.2. According to relevance theory (Sperber and Wilson, 1986/1995; Wilson and Sperber, 1988b), any representation of propositional form can be used in two ways: either *descriptively*, where the representation is used as a truth-conditional description of external circumstances, or *interpretively*, in which case the representation represents another representation with a propositional form which it resembles in content (i.e. with which it shares logical and contextual implications). This dichotomy corresponds to another distinction which is sometimes made between two sorts of propositional attitude (Wilson, 1993; cf. Sperber, 1994a, b, 1996, 1997). Descriptive attitudes are attitudes towards propositions which are regarded as truth-conditional representations of states of affairs: for instance, fearing, demanding or regretting are cases of descriptive attitudes. Interpretive attitudes, on the other hand, are attitudes towards propositions *qua* propositions, i.e. abstract representations which can be entertained as elements of thinking episodes in an agent's mental life: doubting, proposing or wondering are examples of interpretive attitudes. Depending on their semantic content, propositional-attitude and other predicates pick out a specific use of the propositions under their scope. This becomes evident in the two possible uses of the *that*-clause in (79):

(79) a. That the cabinet is corrupt is very sad.
b. That the cabinet is corrupt is completely unfounded.

In (79a), *that* introduces an assumption which is put forth as a description of an actual state of affairs; it is this state of affairs that the speaker describes as very sad. In (79b), however, *that* introduces an assumption which is put forth as a representation of an abstract hypothesis, possibly entertained by some people with various degrees of endorsement; the attitude of the speaker focuses on the abstract hypothesis (rather than the state of affairs represented by it).

The interpretive use of representations couched in propositional form rests on our ability to entertain and manipulate second-order representations in the language of thought. This ability may, in fact, consist of separate specialised systems of metarepresentation. One possibility would be to recognise three such sub-systems, each dealing with a separate sub-type of interpretive use: a *metacommunicative,* a *metalogical* and a *metacognitive* one (see Sperber, 1997). The metacommunicative system would handle representations of utterances, i.e. linguistically communicated propositional forms, originally produced by the speaker (or by another source) at a different time. The metalogical system would be responsible for checking representations for logical consistency, detecting contradictions, and (in more advanced forms) judging a line of argument as valid or undecidable, reasoning about hypothetical possibilities, employing *reductio ad absurdum* and, generally, generating possibilities that have not been specified in advance (cf. Moshman, 1990; Sophian and Somerville, 1988; Byrnes and Overton, 1986). As for the metacognitive system, it would deal with representations of mental states, such as beliefs and desires, and with the human capacity for reflecting on mental states, either one's own or the projected mental states of someone else.

Being varieties of interpretive use, the metacommunicative, metalogical and metacognitive operations share the property of dealing with representations of propositional content, rather than form. It is conceivable that there is a different sub-system for the metarepresentation of linguistic form (covering, for instance, examples of linguistic mention); I will come back to this in chapter 3. For my immediate purposes, I will concentrate on the workings of metacognition.

The metacognitive capacity has in recent years been associated with the so-called *theory of mind* hypothesis in psychology and the philosophy of mind (cf. the contributions in Carruthers and Smith, 1996, and chapter 4). This hypothesis entails precisely that part of the human cognitive mechanism is the ability to know one's own mind as such, i.e. to reflect on one's mental contents and processes and to accommodate the results in a coherent picture of the mental world (see Gopnik and Wellman, 1994; Leslie, 1995, for different views on the architecture of theory of mind). The theory of mind gives a domain-specific, causal-explanatory framework to account for/predict phenomena in the mental domain.

I now come to the core of my proposal: There is an obvious sense in which expressions of epistemic modality fit into a representational model of the mind. In their epistemic uses, modals like *may, must, should,* and *ought to* communicate a logical relation between a certain proposition and the speaker's belief-set. From the speaker's point of view, the employment of epistemic modality rests crucially on the ability to reflect on the content of one's own beliefs, to take into account the reliability of those beliefs (i.e. the relative strength with which they are entertained),[28] and to perform deductive operations on them. The above processes jointly presuppose the ability to conceive of one's mental contents as representations distinct from reality which may bear a variable degree of correspondence to the actual world. From the hearer's point of view, the comprehension of an epistemically modalised utterance involves making reference to the communicator's evaluation of a certain proposition in terms of mentally represented evidence.

On this picture, epistemic interpretations of modal verbs involve a propositional representation being used interpretively: the complement of the verb (:the embedded proposition) is not used as a truth-conditional representation of a state of affairs in the external world but as a representation of an abstract hypothesis, which is considered to be compatible with/entailed by the speaker's set of beliefs. This is the reason why epistemic interpretations are typically taken to convey how much evidence the speaker possesses for the embedded proposition. This type of modal interpretation contrasts with root modal interpretations, where modality operates over propositions handled as truth-conditional descriptions of states of affairs (in the actual, or in an alternative - ideal, stereotypical, etc. - world). Epistemic operators take scope over propositions which are entertained and manipulated *qua* propositions by the speaker. To put it differently, the proposition embedded under an epistemically understood modal is not to be treated as directly picking out a state of affairs in the world, but as describing what Sweetser (1990) has called an ' epistemic object'.

If correct, the metarepresentation hypothesis about epistemic modality will afford a number of predictions about the truth-conditional behaviour of epistemic modal

[28] *Strength* is defined as the cognitive counterpart of the philosophical notion of subjective probability: i.e. a non-representational property which captures the degree of confirmation/support assigned to any given stored assumption and is determined by the assumption's initial formation and subsequent involvement in various cognitive processes. I here follow Sperber and Wilson (1986/1995: 76ff.).

interpretations, their contribution to relevance, and their acquisitional peculiarities. I will comment on these aspects of epistemic modality in later chapters. For the present, I will illustrate how the metarepresentation hypothesis fits with the account of modality already developed for root interpretations in the previous sections. But first I wish to remove a possible objection to the metarepresentational account which might arise at this stage.

Isn't it the case, one might ask, that virtually all assertions are backed up to various degrees by the speaker's belief-set, and can thus be considered as conclusions based on internally represented evidence? This is, after all, what underlies Grice's second maxim of Quality: 'Do not say that for which you lack adequate evidence' (Grice, 1975: 67). Does this mean that we should mark all regular assertions as cases of interpretive use? Obviously, such a move would trivialise what the metarepresentational account is trying to capture: namely, the fact that the speaker has chosen to mark the proposition embedded under the modal as a conclusion and to linguistically indicate the strength of this conclusion by the conceptual information encoded by the modal verb (including both the sort of logical relation and the type of evidence involved in the inferential process). Naturally, assertions come with various degrees of speaker commitment which are often left implicit. In (80), for instance, the speaker may use intonational and other means to convey that her utterance is supported by all the evidence she possesses, although she cannot subscribe wholly to its truth because she manifestly still lacks some crucial information:

(80) Judy likes caviar.

In a case like (80), the strength ascribed to the proposition expressed by the utterance will be wholly pragmatically inferred, rather than computed on the basis of a linguistic trigger. By contrast, modal expressions, on their epistemic interpretations, typically mark the proposition embedded under them as a conclusion. In the specific case of English modal verbs, the conclusion is motivated in terms of inference. Modal expressions in other languages may motivate a conclusion in terms of evidence from perception or from communication (for examples, see Palmer, 1986 on modality; cf. also the evidential adverbs *apparently*, *clearly*, and *allegedly*, *reportedly*).

2.4.2 Derivation of Epistemic Interpretations

Consider the following examples:

(81) Brian's resignation may prove a big mistake.

(82) Some of the neighbours must have seen the burglars.

I will examine each utterance in turn. On hearing (81), the addressee will have to determine the domain of propositions which is compatible with the embedded proposition p. Since p involves a future event, the speaker cannot simply reason from facts when evaluating the proposition in the present. That is, the speaker cannot trust the background assumptions she uses to evaluate p to accurately and fully match states of affairs in the future. Therefore, the speaker should be taken to communicate that p describes a possible conclusion with respect to a set of assumptions available specifically to her (and, although nothing in the speaker's beliefs excludes the possibility that Brian's resignation proves a mistake, there may be circumstances in the world which actually will). The resulting interpretation of (81) is, naturally, epistemic.

Example (82) is very similar to (81). The embedded proposition (let's call it q) has a determinate truth-value at present, since it refers to an event which either has or has not taken place in the past. The speaker cannot have intended a factual propositional domain as the value of D: if she had access to all the facts, and they jointly entailed q, she could have uttered the unmodalised counterpart of (82). On the most plausible interpretation of the utterance, the speaker lacks complete knowledge of what happened at the relevant time-slot in the past (i.e. during the burglary). All she can do, therefore, is reason on the basis of incomplete and partly-supported evidence which she reconstructs from both general encyclopedic and situation-specific information about burglaries (e.g. that the burglars have used one of the usual methods of getting into the house, that they were exposed at least some of the time, that the neighbours pay some attention to what takes place in nearby properties, etc.). Thus, q is presented as a necessary conclusion given an epistemic background domain.

The enriched logical forms of the examples are given in (81')-(82'):

(81') $[_p$[Brian's resignation proves a big mistake]] is compatible with $D_{bel.}$

(82') $[_q$[Some of the neighbours have seen the burglars]] is entailed by $D_{bel.}$
(D_{bel} is the domain consisting of the speaker's beliefs;
double brackets indicate metarepresentation)

This account captures Kratzer's insight that epistemic modality and root modality 'involve a different categorisation of the facts' (Kratzer, 1981a: 52). In root interpretations, the modal domain includes propositions taken as descriptions of states of affairs; in epistemic interpretations, the modal domain D is relativised to a set of propositions which form part of the speaker's belief-set - and thus participate in her mental life. This is not to say that beliefs themselves are not truth-conditional descriptions of states of affairs external to the individual. However, when they estimate epistemic possibility or necessity, humans are not simply concerned with such transparent properties of internally represented propositions, but focus rather on the representational properties (such as accuracy or completeness) of their belief-system as such. In cases where the speaker has full cognitive access to assumptions that can determine the truth/falsity of the proposition embedded in the modal, there would be no need to relativise the modal restrictor to the domain of her beliefs. She could simply convey that the embedded proposition holds or is ruled out by objective circumstances or facts in the external world.

The last point can be used to explain certain puzzles about the relation between an assertion containing epistemic *must* and a non-modalised assertion. As Lyons (1977: 808) observes, 'although it might appear that a statement is strengthened by putting the proposition that it expresses within the scope of an operator of epistemic necessity, this is not so, as far as the everyday use of the language is concerned'. He remarks that an utterance such as (83a) is felt to be stronger than (83b) and to convey a higher degree of speaker commitment:

(83) a. San Marino is the country with the highest life expectancy in the world.
b. San Marino must be the country with the highest life expectancy in the world.

Contrasts in the *must-is* interface become more apparent in observational utterances, i.e. utterances that express propositions verifiable through or directly reporting facts from perception. Assume that Peter is watching a football match. At some point the ball is kicked off the field. Peter may produce either of the responses in (84) but he will be taken to be recommending a higher degree of commitment to the proposition expressed by (84a) than to the proposition embedded under *must* in (84b):

(84) a. That was an off-side.
b. That must have been an off-side.

I propose that the reason for the asymmetry is that (84a) is presented as a factual assumption, guaranteed by Peter's (uninhibited) perceptual access to his environment. Perceptual beliefs, although not necessarily more likely to be true, are normally assumed to be causally related to the structure of reality; therefore, they are considered to be our securest form of contact with the world around us (Dancy, 1985: 178). Since we trust our perceptual experience to deliver information of high epistemological respectability, it follows that other sources of knowledge (e.g. inference) will be valued less when it comes to the assessment of the same piece of information.

A similar point applies to the examples in (83a) and (83b): again only the first is put forward as a piece of factual information, i.e. an assumption which the speaker trusts to be a true (or potentially true) description of a state of affairs in the world. What is more, on my analysis, epistemically interpreted modal verbs involve the metarepresentation of an assumption which is evaluated on the basis of evidence available to the speaker. So, although (83b) conveys that the speaker possesses compelling evidence about the country with the highest life expectancy, the possibility is left open for evidence lying beyond the speaker's beliefs to disconfirm the embedded proposition. This is what underlies the intuition that (83b) is weaker: by bothering to modify her proposition with an epistemic modal operator, the speaker explicitly indicates that she does not want to communicate the full range of effects the unmodalised utterance in (83a) would cause, or at least that she wants to moderate their strength.

On a relevance-theoretic account, (83b) is actually expected to produce extra or different cognitive effects in comparison to (83a) simply because it is both structurally and semantically more complex, and thus more costly in processing effort. The same prediction arises in other epistemic environments, which also require the setting up of a metarepresentation. Compare:

(85) a. There are nine planets in our solar system.
b. As far as I know/In my view, there are nine planets in our solar system.

The metarepresentational analysis extends quite easily to the other modals. Consider (86):

(86) That should be the plumber. (on hearing the doorbell ring)

The speaker cannot be certain that the circumstances in the actual world guarantee that p is true. The only evidence she has for the embedded proposition comes from her beliefs about the normal/expected course of events. According to those beliefs, the plumber is expected to arrive some time after the speaker called him. So if the circumstances in the world have developed as the speaker expects them to, it follows that whoever is ringing the bell is indeed the plumber. This explains why notions such as 'prediction', 'expectation' or 'probability' have been assumed to feature in the semantic entry of *should* (see Ehrman, 1966; Walton, 1988; Bybee *et al.*, 1994 respectively). According to what the verb conveys, if reality develops in accordance with the speaker's expectations, the proposition embedded under the modal will turn out to be true (although it could still be contradicted by evidence which at present lies outside the speaker's cognitive environment).

This line of argument affords an explanation of the subtle differences between *must* and *should* when they function as epistemic markers. Consider the following pair:

(87) a. John must be easy to talk to.
b. John should be easy to talk to.

R. Lakoff, discussing these examples, has observed that they are appropriate in different circumstances (Lakoff, 1972). Suppose that speaker and hearer are standing outside John's office. (87a) could be uttered if there were indications that John had a visitor and they were both having a good time. By contrast, (87b) would be acceptable if the speaker knew nothing about John's habits from first-hand experience, but had heard that in principle John was very kind with his students. Lakoff concludes that *should* is used in the case of a likelihood based on future expectation and verifiable in the future, and *must* of a likelihood based on present conjecture and verifiable in the present.

The interpretations which Lakoff offers for (87) unfold quite easily from the proposed semantics for the two modals. Although both utterances are guesses based on partial evidence (and hence involve epistemic readings of the modals), each verb allows for a different restrictor, and, therefore, a different type of evidence. *Must* leaves open the type of evidence that supports the embedded proposition $_p$[John is easy to talk to]. In the situation Lakoff describes, the most accessible type of evidence capable of causing an adequate range of cognitive effects is John's behaviour with his visitor. *Should*, on the other hand, admits only evidence which is based on norms/expectation. Consequently, (87b) is appropriate when there is no direct experience of John's behaviour. The type of admissible evidence also bears on the verifiability of each utterance. In the case of perceived evidence, as in *must*, the embedded proposition p is verifiable at present, whereas expectation-based evidence, as in *should*, needs to await support or disconfirmation from future states of affairs.[29]

I should note here that my analysis of *should* does not affect in any way the monosemy argument I have been advocating for the modals. On a first pass, one might be worried about the possibility of giving both root and epistemic interpretations to a verb which semantically admits only a stereotypical or normative restrictor. This is not a real problem, however, since it is precisely that semantic information which allows for both types of interpretation in *should*: root interpretations occur when normative assumptions are regarded as representations of external states of affairs, whereas epistemic interpretations arise when the expectation-conforming evidence is focused upon *qua* set of internal propositional representations. In fact, the possibility of interpreting *should* as expressing both root and epistemic (weak) necessity while retaining the feature of stereotypicality rather favours a unitary meaning approach over, say, an ambiguity view. On the ambiguity position, the verb would be ascribed two unrelated clusters of conceptual content, one having to do with social normativity (root interpretation) and the other with likelihood (epistemic interpretation - see Ehrman, 1966 for an explicit adoption of this view). By contrast, the monosemy approach coupled with a metarepresentational stance on epistemic modality can naturally accommodate the fact that the specification of stereotypicality will 'percolate' from external facts to internal evidence for the embedded proposition.

A similar possibility arises with *ought to*, where the semantically specified domain is that of ideals, moral imperatives and the like:

(88) This problem ought to be very easy for a mathematical genius like you.

What about *can*? My semantic analysis precludes any epistemic uses, since the verb semantically restricts the value of admissible modal domains to the factual one. This restriction seems to receive support from the impossibility of constructing utterances where *can* is used instead of epistemic *may*:

---

[29] The present account also explains the difference in acceptability between (i) and (ii) noted in Quirk *et al.* (1985, §4.56):

(i) There should be another upturn in sales shortly.

(ii) ?There should be another disaster shortly.

Unlike *must*, *should* (and *ought to*) encode normative concepts: in (ii), it is odd to suggest that the occurrence of a disaster follows from our expectations about the normal/prescribed course of events. Quirk *et al.* suggest that the epistemic readings retain from the root readings the implication that the embedded proposition is desirable; (ii) is odd because it somehow suggests that the speaker favours disasters. This explanation is compatible with mine, in so far as what is normatively prescribed is also normally desirable.

(89) a. He may/?can have been joking when he said that.
b. Michael may/?can well get his degree next year.

(90) Do you think that James is hiding something from the authorities?
a. He may be, and then again he may not.
b. ?He can be, and then again he cannot.

The intuition that *may* and *can* are typically used to communicate two distinct types of possibility has been expressed by several authors. Leech (1987), for instance, argues that *may* expresses factual possibility and *can* theoretical possibility; van der Auwera (1986) takes *may* to convey indeterminacy and *can* contingency (see also R. Lakoff, 1972; Bolinger, 1989). What these authors standardly emphasise is the contrast between epistemic interpretations of *may* and root interpretations of *can,* as well as the fact that *can* does not accept epistemic interpretations.[30] Let me go back to (90). In (90a), the speaker makes manifest that the proposition $_p$[James is hiding something from the authorities] is compatible with the set of her beliefs, and goes on (after some deliberation) to assert that the negation of this proposition (~p) is also compatible with her beliefs.[31] To put it differently, the speaker's current beliefs leave open the possibility for either p or ~p to come out as true: there is nothing unacceptable or ungrammatical here. By contrast, in (90b) the speaker communicates that p is compatible with the set of factual propositions - hence that the state of affairs in the actual world allows for p to be the case - and goes on to suggest that p is *not* compatible with the set of factual propositions - thereby creating a contradiction.

Using the root-epistemic contrast, we can also explain why (91a) seems to rely much more on the future for the verification of the embedded proposition than (91b):

(91) a. James may be hiding something.
b. James can be hiding something.

In the former, but not in the latter example, we expect the speaker's knowledge to develop so as to judge p as true or false. Similarly, we can account for the fact that in (92) it is the utterance containing *may* which expresses the greater likelihood that James is hiding something:

[30] *May* does, of course, accept root interpretations, as I have shown in section 2.3.1.
[31] On the scope of negation in epistemic modal interpretations, see section 3.1 below.

(92) James can be hiding something from the authorities; indeed, he MAY be hiding something from them.

The value of the restrictor in *can* being what it is, no percolation effects can be used to elicit epistemic interpretations: these would violate the semantics of the verb. The presence of lexical-semantic factors which guarantee the non-convergence of *may* and *can* might be viewed as a means of securing a division of labour between the two lexical items - a synchronic situation which has gone through various stages of partial overlap of the meanings of the two modals (Tanaka, 1990; Traugott, 1989; Bybee *et al.*, 1994). The pressure for differentiation between the two verbs is probably also responsible for the relatively low frequency of simple root possibility interpretations for *may* (since *can* grammaticalises precisely this part of the conceptual space of modality in English).

Still, one might argue, there are certain environments which can be said to elicit epistemic interpretations of *can* - specifically, interrogative, negative and 'generic' environments (Sweetser, 1990; Walton, 1988):

(93) a. Can this allegation (ever) be true?
b. This allegation can't be right.
c. Defamatory allegations can bring an end to a political career.

If these are truly epistemic interpretations of *can*, then the semantics I have proposed for *can* will turn out to be incorrect. However, I do not think there is any compelling reason to attach epistemic readings to the utterances in (93). The usual factual readings will do. Support for this position can be adduced from the behaviour of epistemic modal readings in *yes-no* interrogatives and negative environments in English. Anticipating the discussion in 3.1, let me note that interrogative-initial modal verbs cannot be interpreted epistemically in genuine requests for information: the speaker cannot be expecting the addressee to supply the information whether a proposition is compatible with/entailed by the speaker's beliefs. Root readings, by contrast, do not impose such constraints on the type of interrogative they occur in. Since (93a) may be a well-formed request for information as well as a deliberative question, it follows that *can* here is interpreted as a root verb. Likewise, if *can't* in (93b) is indeed an epistemic, this will be the only case in English where an epistemic reading of a modal verb falls in the scope of negation. So despite the traditional classification of *can't* as a suppletive form for internal epistemic negation ('It is epistemically necessary that not'), I think it is reasonable to take the verb as having its usual root reading (thereby communicating the stronger 'It is factually necessary that not').[32] Finally, so-called generic uses of *can* in (93c) do not involve epistemicity but rather a generalisation over factual instances, i.e. a habitual reading. Other similar uses of *can* are given in (94):

(94) a. A good football player can be short.
b. Italians can be good cooks.

I conclude that the examples in (93) pose no threat to the overall validity of the claim that *can* encodes a factuality restrictor.

Let me sum up my position on epistemic modality. Epistemic interpretations of modals arise in contexts in which it is mutually manifest that an individual, in drawing a conclusion, say p, is not in a position to take into account every proposition that could affect the truth of p, because he/she is ignorant of existing evidence. The individual will then draw the most justifiable conclusion according to presently available evidence, and the overall degree of support for it will depend on the degree of strength of the various premises. Stated differently, the individual ends up with a logical conclusion whose premises are nevertheless more or less disputable, so that it is ultimately ascribed a correspondingly low degree of strength. What sets epistemic interpretations apart from other modal interpretations is that both the proposition embedded under the modal and the evidence for it are metarepresentational assumptions capturing the individual's internal representation of reality, which are likely to evolve and be revised, as new evidence becomes available.

2.4.3 Derivation of Alethic and 'Objective Epistemic' Interpretations

I now wish to compare epistemic modality with another type of modal concept, which logicians have termed *alethic* (or logical) modality (McCawley, 1981; Lyons, 1977). A proposition is a logical necessity if it is entailed by the set of propositions in any domain D. A proposition is a logical possibility if it is compatible with the set of propositions in any domain D. The following are examples of logical possibility and necessity:

(95) a. I haven't won the Lottery yet, although in theory I could have.
b. It is possible that someone I know is going to win the Lottery.
c. Winning the Lottery does not necessarily make one happy.

---

[32] Another solution would be to insert *can't* as a separate lexical item in the lexicon.

Several linguistic discussions ignore alethic modality. Lyons' is a rather typical remark when he notes that alethic uses are the products of a 'rather sophisticated and impersonal process which plays little part in ordinary non-scientific discourse' and is 'secondary in the acquisition of language' (Lyons, 1977: 845, 849). This is especially true with regard to the modal verbs in English (and in other languages). The alethic interpretation of *must* in (96) is, indeed, felt to be contrived:

(96) He is a bachelor, so he must be unmarried.

Nevertheless, since the boundaries between alethic and epistemic modal interpretations are often empirically hard to distinguish, there have been attempts at conflating the two types of modality. One line of argument has been to reduce alethic interpretations to epistemic interpretations, since they are not morphologically distinct in English, or, apparently, any other language (Palmer, 1986: 11). The opposite suggestion has also been made: some writers have proposed that epistemic modality can (and should) be reanalysed as a subtype of alethic/absolute modality (Kneale and Kneale, 1962: 93; Hughes and Cresswell, 1968: 27; cf. Karttunen, 1972: 14ff.). On this view, we could interpret (97) as conveying that it is absolutely/logically necessary that, if the premises are true, then the conclusion is also true:

(97) If it snows throughout February, my birthday must be a snowy day. (assuming the speaker's birthday is on the 11th February)[33]

There are good arguments against conflating epistemic and logical modality (Karttunen, 1972; cf. Iatridou, 1990). First, alethic *must p* is stronger than the unmodalised proposition p, since the former means that p is not just true but necessarily true (true in all possible worlds). Epistemic *must p*, though, is felt to be weaker than p, for reasons explained in previous paragraphs.[34] Second, in modal logic p and *possibly ~p* are consistent. However, in natural language there is some incompatibility between the two clauses of:

(98) ?It isn't raining in Chicago, but it may be raining there.

---

[33] Example adapted from Hughes and Cresswell (1968: 27).
[34] This is why we can infer p from alethic *must p* but not from epistemic *must p*.

Third, logically speaking, the form *must* $(p \rightarrow q)$ makes a different statement from $p \rightarrow$ *(must* q*)*. Yet there is little difference between the English forms in (99):

(99) a. It must be that, if Bill has a diamond ring, he has stolen it from someone.
b. If Bill has a diamond ring, he must have stolen it from someone.

As Karttunen (1972) has rightly observed, there is also a basic conceptual difference between the two types of modality. In my terms, the sort of assumptions that will be included in the modal restrictor differs in alethic and epistemic modality. As a result, it is a mistake to think that the one can be made to collapse into the other.

One thing these analyses *were* right about was that alethic and epistemic modality bear a close relation. In fact, I want to claim that the ability to entertain alethic concepts is essentially of the same type as the ability to employ epistemic notions, insofar as they both involve an aspect of metarepresentation.

While discussing the facets of the human metarepresentational device in an earlier section, I mentioned metalogical activity as part of it. Metalogical processes consist in the subject's 'displaying' propositions as abstract hypotheses and (mostly consciously) performing deductive operations on them. Metalogical reasoning rests on the assumption that logical relations such as entailment, compatibility, contradiction and so on, obtain between propositional contents in the abstract, mind-independently. The interest in performing metalogical computations, then, lies in recovering these logical relations. Alethic/logical modality presupposes precisely an ability to reason about what is simply possible or necessary, thereby considering alternatives which are not included among presently available circumstances in the real world, or states of affairs described by the agent's current beliefs, but predicted by general logical laws.

What alethic modality shares with epistemic modality, then, is the fact that they are both instances of the interpretive use of propositions. They both rest on the capacity to envisage propositional representations as *representations* which enter into specific logical relations. Alethic concepts differ, however, in that they involve propositions viewed as mind-independent abstract entities, whereas epistemic concepts relate propositions which are treated as participants in the agent's mental life (and therefore require a fully set-up theory of mind).

Within the lexical-grammatical system of English modal expressions I have described, alethic interpretations of modals will come out as yet another possible enrichment of the underlying basic meaning of modals (in cases where it is permitted by the specified modal restrictor). It is true that the distribution of such interpretations is very restricted, especially as far as the English modal verbs are concerned. They seem to occur more freely with modal adverbs and adjectives as below:

(100) a. It is possible that there is life on other planets.
b. One in three necessarily wins.

The metarepresentational analysis of both alethic and epistemic modal concepts entails the following prediction: In case the logical domain and the epistemic domain turn out to be extensionally equivalent (i.e. if the speaker's knowledge spans all logical possibilities), then it will be impossible to differentiate between a logical and an epistemic interpretation of a modal expression. Such mixed interpretations do indeed occur. Lyons (1977) offers some examples and tentatively proposes that they belong to a separate category intermediate between epistemic and alethic modality. Imagine that a murder has occurred in a secluded hotel, where there is no communication with the outside world. The murderer then has to be one of the ten residents, who immediately become suspect and have their alibis checked by the police. After finding out that all of these suspects have unassailable alibis except for the guest in room 31, the police inspector produces (101):

(101) The guest in room 31 must be guilty.

Lyons (1977: 798) analyses similar utterances as cases of *objective epistemic modality*, where a conclusion reached by the speaker presents itself as an objective fact. He contrasts these with classic examples of epistemic modality, which is termed *subjective* and is thought to be a product of 'opinion, or hearsay, or tentative inference' (Lyons, 1977: 799). Apart from objective epistemic necessity, we find cases of objective epistemic possibility. In the same detective-story context, and before the investigation began, the inspector might utter:

(102) The guest in room 31 may be guilty.

Since this guest is among the ten residents in the hotel at the time of the murder, there is an objective possibility that he is the murderer. As Lyons puts it (1977: 798):

> The speaker might reasonably say that he knows, and does not really think or believe, that there is a possibility (and in this case a quantifiable possibility) of

> [this guest's being guilty]; and if he is irrational, his own subjective commitment to the truth or falsity of the proposition '[The guest in room 31 is guilty]' might be quite unrelated to his knowledge of the objective possibility, or degree of probability [1/10], of its truth, in the way that a gambler's subjective commitment to the probability of a particular number coming up in roulette might be quite unrelated to the objective probabilities.

That this is not really a distinct type of modal interpretation, but rather a conflation of two interpretations, can be argued for as follows. In the first place, Lyons' explanation faces a difficulty: the distinction between knowledge and belief, which is crucial for his distinction between objective and subjective epistemic modality, is dubious from a psychological point of view. Evidently, the individual cannot distinguish knowledge and belief on subjective grounds alone. In examples (101) and (102), the individual's knowledge about the circumstances of the murder is - from an internalist perspective - equivalent to a set of strongly held beliefs, coupled with the assumption that there will be no further data which could have a bearing on the truth of these beliefs. This is guaranteed by explicitly stating all factors that could have a bearing on the identity of the murderer. The speaker can therefore be sure that no additional evidence will influence the truth of her statement about the guilt of the guest in room 31, so that her statement will be fully backed up by her cognitive environment. What appears to be an objective sort of epistemic modality is thus a conflated type of interpretation, where ordinary epistemic necessity or possibility is evaluated with respect to a fully determinate and strongly supported body of evidence. What lies behind Lyons' intuition that this interpretation is close to alethic/logical modality is precisely the fact that the set of circumstances considered as evidence here span the range of logically possible circumstances. By being mutually manifest to the inspector and his interlocutor, this fact explains why (101), for instance, is perceived as a stronger statement than its unmodalised counterpart in (103):

(103) The guest in room 31 is guilty.

2.5 Concluding Remarks

My aim in this chapter has been to strike a middle way between polysemy-based and radical monosemy accounts of English modals. I have tried to offer a semantics rich enough to allow for differences in content in the various modals, and yet underspecified to the extent of drawing on extensive pragmatic inferencing until it yields a complete truth-evaluable representation (:the proposition expressed by the modal utterance). I have attempted to draw together root and epistemic interpretations and to demonstrate how they can arise from what is - for most verbs - a common basic meaning. Still, epistemic interpretations came out as significantly different from root ones: I have argued that the ability to employ epistemic concepts is part and parcel of human theory-of-mind abilities, and therefore epistemic interpretations of modals make use of metarepresentations in a way that root interpretations do not.

I have not yet fully cashed out what I have called the 'metarepresentation hypothesis' for epistemic modals. In the following chapters, I go back to the types of evidence standardly adduced to support the existence of polysemy in English modals and attempt a reanalysis in terms of the metarepresentation hypothesis.

3

Removing Objections to a Unitary Semantic Analysis

There is nothing like looking, if you want to find something...
You certainly usually find something, if you look,
but it is not always quite the something you were after.
J. R. Tolkien, *The Hobbit*

3.0 Introductory Remarks

The goal of the previous chapter was to provide arguments for a monosemy-based account of the English modal auxiliaries and to sketch a plausible pragmatic derivation of their various contextual interpretations. My next aim is to revisit two arguments which are widely considered to provide compelling evidence for the grammatical status of the root-epistemic distinction. The first is based on the isolation of specific clusters of grammatical features, whose distribution seems to uniquely determine either an epistemic or a root interpretation. The second includes tests which are meant to demonstrate the non-truth-conditional nature of epistemic modality markers. To these I will add a third point: the postulation by some authors of a distinct category for speech-act modality, set up as a natural extension of a polysemy-based account of modality. All three arguments are supposed to pose a challenge for any approach to the modals which assumes a 'flat', unitary semantics. Such approaches, the claim goes, are incapable of distinguishing between root and epistemic interpretations in terms of distribution or truth-conditional behaviour; moreover, they would be at a loss to fit speech-act modality into an already overloaded pragmatic machinery.

What I propose to do in this chapter is offer an alternative explanation for the data invoked by each of the above arguments. I will firstly claim that, far from being ill-adapted to capturing the differences between the two major classes of modal interpretations, the unitary semantic analysis put forth in the previous chapter can naturally accommodate them. Furthermore, the analysis can explain inconsistencies in distribution and truth-conditional behaviour, which would be problematic on the alternative account. Finally, I will show how the metarepresentation hypothesis can deal with the category of speech-act modality in an economical and theoretically attractive way. If this analysis is right, then the areas of distribution, truth conditions and speech-act modality are indeed testing grounds for semantic/pragmatic theories of modality, but they do not demonstrate the superiority of lexical polysemy over other semantic alternatives as they have been claimed to do.

I will devote one section to each of the three arguments. Section 3.1 is concerned with the (alleged) grammatical reflexes of the root-epistemic distinction. Section 3.2 is a discussion of the truth-conditional behaviour of modals. The final section (3.3) addresses 'speech-act' modality.

3.1 Syntax and Interpretation

3.1.1 Root-Epistemic Contrasts

3.1.1.1. Background. In the generative literature, there is a long line of research which assigns distinct structural features to root and epistemic readings of modal verbs. In early work (Ross, 1969; Perlmutter, 1971; Huddleston, 1974), which assumed a main verb analysis of modals, root modals were considered transitive, whereas epistemic modals were considered intransitive.[1] Later, the root-epistemic distinction was related to the raising-control alternation (Jackendoff, 1972). Consider the utterance in (1):

(1) The children may sleep outside.

[1] This approach is now obsolete: there are many reasons for treating English modals as distinct from main verbs. For instance, modals do not undergo number agreement; they do not appear together; they lack gerunds and infinitives; they undergo subject-auxiliary inversion; they precede *not*; finally, they block *do*-support (cf. Jackendoff, 1972: 100-1). Interestingly, these properties hold regardless of whether modals belong to the root or the epistemic interpretive class.

According to Jackendoff, (1) on its epistemic reading corresponds to a raising structure of the type in (2): the logical subject of the verb seems to be the whole proposition in the modal complement. The utterance in (1) on its root reading is taken to correspond to a control structure of the type in (3): it is the children that are allowed to sleep outside:

(2) a. The children seem to sleep outside.
b. The children appear to sleep outside.
c. The children happen to sleep outside.

(3) a. The children try to sleep outside.
b. The children refuse to sleep outside.
c. The children agree to sleep outside.

Epistemic interpretations thus come out as one-place operators which take the interpretation of the complement of the modal as an argument. By contrast, root interpretations are treated as two-place operators which take the interpretation of the modal complement and a certain NP (typically the surface subject of the complement) as arguments. The idea behind Jackendoff's proposal is that root readings involve (required or permissible) actions performed by an agent, while epistemic readings involve the evaluation of a proposition as possible or necessary. The parallel with raising-control structures is made evident in the possible paraphrases of the epistemic and root reading of (1) in (4) and (5):

(4) a. The children are likely to sleep outside. (epistemic)
b. The children are obliged to sleep outside. (root)

(5) a. It is possible that the children will sleep outside. (epistemic)
b. It is possible for the children to sleep outside. (root)

Since Jackendoff (1972),[2] there have been many attempts to separate the two interpretive classes of modality on syntactic grounds (often building on older observations about certain distributional differences between root and epistemic readings). An assumption shared by most syntactic approaches is that the epistemic-root contrast is co-extensive with a distinction between high- or low-scope modals. Moreover, it is standardly thought

[2] Jackendoff intended the root-epistemic split to be treated as a semantic, rather than a syntactic, distinction (alongside the raising-control split).

that the complements of root and epistemic readings possess distinct grammatical features. In what follows, I present a summary of the arguments for a structural root-epistemic split (see Hofmann, 1966; Boyd and Thorne, 1969; Ross, 1969; Perlmutter, 1971; Jackendoff, 1972; Huddleston, 1974; Steedman, 1977; Zubizarreta, 1982; Coates, 1983; Roberts, 1985; Palmer, 1990, 1995; Heine, 1995; Adger, 1997; Cinque, 1999).

3.1.1.2 Negation. It is often pointed out that epistemic modals, unlike root modals, always scope outside negation:

(6) a. You may not be given this opportunity again. (epistemic)
' It is possible that you will not be given... '.
*' It is not possible that you will be given... '.
b. You may not enter. (root)
'You are not allowed to enter'.
*'You are allowed not to enter'.

According to Coates (1983: 237-9), 'negation affects the modal predication if the modal has Root meaning', while '[it] affects the main predication if the modal has Epistemic meaning'. An exception cited is root *must*, which also scopes over negation. In order to negate root necessity, the suppletive form in (8) is used:

(7) You must not be late.
'You are required not to be late'.
*'You are not required to be late'.

(8) You need not be late.

3.1.1.3 Ordering Constraints. When an epistemic and a root modal occur in adjacent positions, the epistemic modal always has higher scope than the root modal. In English, there is a syntactic constraint, which disallows the co-occurrence of two modal verbs, so I have used semi-modals and other modal constructions to illustrate:

(9) epistemic > root
a. They may have to go soon.
b. Brian must be able to find a ticket.
c. They might have to stay here.

(10) *root > epistemic

a. *It must be likely to rain tomorrow.

b. *The two formulations ought to have to be equivalent.

In support of this argument, it is noted that the same constraint holds in Hawick Scots, where double or even triple modal verbs are allowed. Consider the following data and glosses from Brown (1991):

(11) a. He should can go tomorrow.
'He ought to be able to go tomorrow'.
b. He would could do it if he tried.
'He would be able to do it if he tried'.
c. He'll might can come in the morn.
'It is possible that he will be able to come in the morn'.
d. He might could have to go.
'It is possible that he was allowed to have to go'.

According to Brown (1991: 76-8), the only possible combinations of modal auxiliaries follow the pattern in (12) with no two consecutive modals belonging to the same class:

(12) future *will* > epistemic > root

More recently, Cinque (1999: 76) in a cross-linguistic investigation of the ordering of functional heads, concludes that the epistemic > root scope constraint belongs to Universal Grammar.

<u>3.1.1.4 Interrogatives</u>. Epistemic modals, unlike root modals, cannot occur sentence-initially in *yes-no* interrogatives:

(13) May the race start?
'Is there permission for the race to start?'
*'Is it possible that the race starts?'

(14) Should John leave?
'Is it required that John leave?'
*'Is it predictable that John will leave?'

3.1.1.5 Properties of the Subject. Root readings, unlike epistemic readings, impose selectional restrictions on the subject of the complex. Utterances with expletive or inanimate subjects in the complement of a root interpretation are anomalous, while similar utterances with epistemic interpretations are fine:

(15) a. There may be a demonstration today. ('It is likely.../*...is allowed')
b. It may be raining. ('It is likely.../*...is allowed')
c. The political uncertainty may lead to early elections. ('It is likely.../*...is allowed')

(16) a. There must be a demonstration today. ('It is certain.../*...is required')
b. It must be raining. ('It is certain.../...*is required')
c. The political uncertainty must lead to early elections. ('It is certain.../*...is required')

This fact is taken to suggest that epistemic interpretations take (or can take) a single argument in their semantics and that this argument is a proposition; recall that this has been the motivation for treating epistemic readings as essentially raising structures. Root interpretations, on this analysis, resemble control structures in that they are theta-role assigners. The last point is reinforced by the observation that root readings undergo a meaning shift in passivisation, while epistemic readings are unaffected:

(17) a. Relatives may visit the students on Mondays.
b. The students may be visited by relatives on Mondays.

(18) a. The home team may win the game.
b. The game may be won by the home team.

Root (17a) differs in meaning from (17b); the first utterance involves the rights of relatives, while the second the rights of the students. By contrast, epistemic (18a) and (18b) do not differ in conveyed content; they both communicate that there is a possibility that the home team will win the game.

3.1.1.6 Properties of the Verbal Complement. It is often noted that root and epistemic modal readings impose different restrictions on the aspectual profile of the verbal complement. Root interpretations disallow the presence of perfective or progressive aspect in the complement. The following utterances possess only an epistemic interpretation:

(19) a. He must have been very ill then.
b. They might have run away.

(20) a. You must be joking.
b. She may still be seeing him.

Similarly, root readings disallow individual-level stative predicates in the complement. These predicates refer to inherent properties of an individual which cannot be changed (under normal circumstances). Consequently, the utterances in (21) possess only an epistemic interpretation:

(21) a. He must be tall.
b. They may be native speakers of Japanese.
c. People in this part of the world may believe in strange gods.
d. He must know the answer.

Root readings accept stage-level stative or activity predicates in their complement. These predicates refer to states of affairs which can be brought about by an individual. In (22) and (23), the root interpretation of the modal is available:

(22) a. He must behave himself.
b. You should be quiet.

(23) a. You may open the window.
b. They ought to find the code.

A consequence of the aspectual restriction is that epistemic readings have present or past orientation. An inspection of the examples in (19)-(21) verifies that the speaker is concerned with evaluating propositions about past or present events. By contrast, root readings have future orientation. In (22)-(23), the speaker is concerned with (permissible or required) activities which are to occur in the future.

To summarise: I have presented a range of arguments which have been considered in the literature as reliable detectors of a scope difference between the two major interpretive classes of modals. According to the discussion in this section, root interpretations are VP-operators, while epistemic interpretations are IP-operators. It follows that, in English, the type of modal interpretation (root or epistemic) can be uniquely read off from syntactic structure. Similar claims have been put forth for other languages. Picallo (1990) proposes that, in Catalan, modals with epistemic reading are constituents of INFL, while modals with root reading and semi-modals are VP adjuncts; Barbiers (1995) argues for a syntactic difference between root and epistemic readings of modals in Dutch.

Nevertheless, there are reasons to doubt that there is such a straightforward correlation between the pragmatics and the syntax of modality in English. My aim in the next section is to show that the arguments which I have reviewed are at best inconclusive, at worst empirically inadequate. I take up each argument in turn.

3.1.2 Counterexamples and Alternatives

3.1.2.1 Negation. On closer inspection, the claim that, in English, root and epistemic modal readings pattern differently with respect to negation turns out to be simply wrong. As the following examples show, *must*, *should* and *ought to* assume wide scope with respect to negation regardless of their interpretation:

(24) a. These children must not be older than three. (epistemic)
'It is certain that these children are not older than three'.
*'It is not certain that these children are older than three'.
b. They must not leave. (root)
'It is required that they do not leave'.
*'It is not required that they leave'.

(25) a. He should not be late; he only went across the road. (epistemic)
'It is predictable that he will not be late'.
*'It is not the case that it is predictable that he will be late'.
b. John should not drink alcohol. (root)
'It is advisable for John not to drink alcohol'.
*'It is not the case that it is advisable for John to drink alcohol'.

(26) a. She ought not to appear before noon. (epistemic)
'It is predictable that she will not appear before noon'.

*'It is not the case that it is predictable that she will appear before noon'.

b. You ought not to come. (root)
'It is required that you do not come'.
*'It is not required that you come'.

I follow Cormack and Smith (forth.) in assuming that *shall*, *will*, and *would* pattern alongside the necessity modals (the reader can check the relevant examples for himself). The same holds for the semi-modal *is to*:

(27) a. There isn't to be another such president. (epistemic)
Necessary [not *p*]
b. Peter isn't to appear in public. (root)
Necessary [not *p*]

There is a second class of modal verbs which consistently fall under negation. This includes *can*, *could*, *need* and *dare*:

(28) George cannot swim.
'George does not have the ability to swim'.

(29) George could not swim.
'George did not have the ability to swim'.

(30) a. You need not resign.
'It is not the case that you need resign'.
b. The proof need not exist.
'It is not the case that the proof need exist'.

(31) You dare not resign.
'It is not the case that you dare resign'.

It follows that the behaviour of English modals with respect to negation does not offer a straightforward argument for syntactically separating root and epistemic interpretations.[3] Rather the scope divide seems to be broadly drawn along the lines of necessity vs.

[3] Perhaps the difficulty of finding appropriate paraphrases for modal interpretations is mainly responsible for the mistakes in the scope facts concerning negation; see Cormack and Smith (forth.), fn.7.

possibility (with necessity operators scoping over negation, and possibility operators under it). This proposal departs from previous accounts in the literature which have tended to assume that epistemics (unlike deontics, etc.) must of necessity fall outside the scope of negation. Hence it is not embarassed by the fact that there are languages in which it is possible to negate epistemic modality (for cross-linguistic data from Danish, German, etc., see Palmer, 1995).

There is one exception to the conclusion that, in English, the scope of modals with respect to negation is independent of their interpretive class. *May/might* do seem to vary in scope depending on whether they are assigned an epistemic or root interpretation. (32) is a modification of the original example in 3.1.1.2:

(32) a. You may/might not be given this opportunity again. (epistemic)
'It is possible that you will not be given... '.
*'It is not possible that you will be given... '.
b. You may/might not enter. (root)
'You are not allowed to enter'.
*'You are allowed not to enter'.

A possible solution would be to create two distinct lexical entries for the two interpretations of these verbs. Since there are otherwise strong arguments against ambiguity of modal words in English, this issue is left open.

I should point out that I am concerned throughout with sentence negation. One may embed low-scope modals (i.e. modals of the *can* class above) under constituent negation, thereby creating additional interpretations:

(33) You may not go to the reception yet.
'You are allowed [not to go to the reception yet]'.

(34) You can not carry a gun.
'You are permitted [not to carry a gun]'.

The availability of such readings is facilitated by distinctive stress patterns:

(35) You CAN not carry a gun.

Constituent negation differs from sentence negation in that the former can be realised only as *not*, while the latter as either *not* or *n't*. Moreover, the two types of negation yield different tag questions: (33) gives rise to (36), while the same utterance with sentential negation produces (37) (for discussion of these and related issues, see Cormack and Smith, forth.):

(36) You can not carry a gun, can't you? Can [not

(37) You can not carry a gun, can you? Not [can

Let me add that, as a number of commentators have noted (Gazdar *et al.*, 1982; Palmer, 1990; Brown, 1991; Williams, 1994), negation sometimes scopes over canonically high-scope modals (i.e. of the *must* class):

(38) A to B: Shouldn't you be home?
'Is it not the case that you should be home?'

(39) You SHOULDN'T be home?

(40) You should be home, shouldn't you?

(41) A: You should be home.
B: No, I shouldn't.

Following Cormack and Smith (forth.), I assume that the above environments involve what relevance theory calls *echoic use* of language (Sperber and Wilson, 1986/1995; cf. section 3.3 below). In (38) and (39), the speaker A holds the assumption/expectation 'B should be home', and wants to know whether this assumption fails to hold. Tags as in (40) are by nature echoic. Last, in (41), B echoes and rejects an assumption previously expressed by A. These special contexts involve importing material from attributed thoughts/utterances; hence the behaviour of negation in these examples should not affect the generalisations about the ordinary scope of negation and modals.

<u>3.1.2.2 Ordering Constraints</u>. Let me move on to the relative ordering of root-epistemic interpretations within a single clause. There are good reasons to assume that the intra-clausal ordering of root and epistemic readings has conceptual, rather than syntactic,

underpinnings. Firstly, there is already an independent, syntactically motivated scope difference among English modal verbs (broadly, necessity vs. possibility modals), as the data from negation in the previous section have shown. It follows that the epistemic > root constraint cannot also be syntactic (Cormack and Smith, forth.). Secondly, the restriction applies to other constructions in English with similar content (including non-modal verbs, modal adjectives and adverbials):

(42) a. I infer that I am allowed to leave.
b. ??I am allowed to infer that I will leave.

(43) a. It is conceivable that it is permissible to leave.
b. ??It is permissible that it is conceivable to leave.

(44) a. You can possibly leave.
$\text{Possibly}_{\text{epistemic}}$ [can_{root}
b. You may necessarily go back to your country.
$\text{May}_{\text{epistemic}}$ [$\text{necessarily}_{\text{root}}$

Thirdly, the restriction is obeyed in other languages by modals which are syntactically main verbs. Consider the following examples from Modern Greek:

(45) Bori na eprepe na figi.
May-3sing.pres. must-3sing.past.subj.leave-3sing.pres.subj.
'It may be that s/he had to leave'.

(46) *Boruse na prepi na efige.
Can-3sing.prt. must-3sing.pres.subj. leave-3sing.pres.subj.

A conceptual explanation of the epistemic > root ordering might build on the fact that the possibility or necessity of an inference cannot depend on what is socially, physically etc. allowed or required. In different terms, whether an assumption is considered to be compatible with/entailed by an individual's beliefs cannot be subject to social requirements, obligation, permission, etc. The converse is quite unproblematic: that some action is required, allowed and so on, can be open to conjecture or can be a necessity in view of one's beliefs.

The general form of this explanation may be extended to account for the vagaries in the interpretation of adjacent modal (and related) expressions within each major interpretive class:

(47) root > root
   a. You must be able to prove your innocence.
   b. The students may be allowed to take the exam home.

(48) epistemic > epistemic
   a. Necessarily, the solution to this problem may be false.
   b. It should be necessary that a proof exists.

As (47a, b) show, when two root modal expressions co-occur, they generally contribute different types of modal relation or modal domain to the interpretation of the utterance. In (47a), both relation and domain differ: *must* communicates necessity with respect to the law, while *be able to* conveys possibility given an individual's abilities. In (47b), the modal relation is the same (possibility) but the modal domain is different; hence, even though both expressions are used to communicate permission, the sources and targets of the permission are taken to differ. For instance, (47b) is felicitously used on the occasion that university regulations permit the instructors to allow the students to take the exam home. As the reader can verify, exactly parallel observations can be made for the epistemic pairs in (48).[4] In sum, then, it seems that the specifics of the admissible modal co-occurrences, both within and across the root and epistemic classes, are to a large part dependent on construal requirements on pairings of modal relations and modal domains.

To conclude, one may combine the requirements of the two independent scope constraints on English modals, the syntactic and the conceptual one, and derive predictions about the relative ordering of root and epistemic modals in discourse. The case of multiple modals in Hawick Scots is relevant here. If both constraints are effective, the only modal which can occur in second position in Scots should be *can* (both low-scope and root). An inspection of the data in Brown (1991) shows that this prediction is borne out.

---

[4] Occasionally, two modal items may express identical modal relations and domains within the same clause. Lyons (1977: 807) calls these 'modally harmonic' combinations:
   (i) Could I possibly leave?
   (ii) It may possibly rain.

3.1.2.3 Interrogatives. I propose that the impossibility of having epistemics at the beginning of *yes-no* interrogatives can also be explained in conceptual terms. More specifically, it is hard to construct a context in which it would be felicitous for the speaker to ask whether a conclusion is possible or necessary with respect to her own set of beliefs. Once such a context becomes available, however, interrogative-initial epistemics become acceptable. Two such cases are given below. (49) involves deliberative questions, in which the speaker addresses a question to herself in an attempt to evaluate the evidence she has for a certain conclusion. (50) includes cases where the speaker is evaluating publicly available, objective evidence for a conclusion. In both cases, the epistemic interpretation is fairly easy to derive:[5]

(49) a. Must John be a liar?
b. Might John be a liar?

(50) a. Must the patient have cancer?
b. Might the books still be at home?[6]

3.1.2.4 Properties of the Subject. I now turn to the generalisations about the subject of root and epistemic complexes. A first thing to notice is that these generalisations are not absolute: expletives and inanimate subjects do not occur exclusively with epistemics:

(51) a. There must be law and order in the country.
b. The table should be ready for dinner before 7.

(51) presents a class of root interpretations found in so-called deontic statements. The existence of such readings makes the parallel between the epistemic-root and the raising-

[5] A fact often mentioned in conjunction with the interrogative data is the impossibility of getting epistemic interpretations in fronted structures of the following sort (Hofmann, 1966; Jackendoff, 1972):
(i) Only three students may/should/must the professor see.
However, it is not clear how this fact could motivate a structural root-epistemic split. In any case, there are counterexamples to the generalisation (Brennan, 1993: 24):
(ii) Your past and your future might I see, gazing into my crystal ball.
(iii) Full fathom five must your father lie, or we would have seen him here.

[6] There is another class of *yes-no* interrogatives which accept epistemic readings of modals in sentence-initial position:
(i) This is what the teacher said, so it must be right.
(ii) Must it now?
These are cases of the echoic use of language (cf. the remarks on negation in 3.1.2.1). I will not consider such examples further here.

control alternations difficult to establish. If anything, these cases show that root readings may give rise to both control-like and raising-like structures.

The same class of examples proves that root readings cannot be taken to project an external theta role: there is no suitable candidate NP in (51). Obviously, in case a modal communicates permission or obligation, there must be an agent responsible for carrying out the relevant activity or bringing about the relevant state. (51b), for instance, can be interpreted as imposing the obligation on the hearer to prepare the table for dinner; this is a case where the identity of the agent is identified pragmatically. Hence the fact that, typically, root interpretations require the presence of a responsible individual (or agent) is explained in terms of the conceptual requirements for successfully issuing permission or obligation (cf. Searle's (1969) felicity conditions).

The proposed approach is compatible with several facts which have been offered as evidence for the raising-control treatment of modals, while avoiding the shortcomings of the raising-control account. Consider the (alleged) meaning shift which root interpretations undergo in passivisation. (52)-(53) repeat the earlier examples in (16)-(17):

(52) a. Relatives may visit the students on Mondays.
b. The students may be visited by relatives on Mondays.

(53) a. The home team may win the game.
b. The game may be won by the home team.

Notice that not all root readings undergo such a meaning shift. Deontic statements involving inanimate subjects are a notable exception (examples from Newmeyer, 1970):

(54) a. Sam must shovel the dirt into the hole.
b. The dirt must be shoveled into the hole by Sam.

(55) a. Visitors may pick flowers.
b. Flowers may be picked by visitors.

I suggest that the perceived meaning change in passivisation, when it occurs, is due to the different contexts made available for processing the active and passive version of an utterance. The difference has to do with what is broadly known as the Figure-Ground, or Subject-Complement distinction. (52a), on its most natural interpretation, conveys something about the relatives, while (52b) is rather taken to communicate something about the students. In ordinary situations of issuing permission or obligation, it matters who the person targeted by the permission or obligation is. Hence the modal domain for the interpretation of (52a) will include assumptions about the relatives, while the modal domain for the interpretation of (52b) will involve assumptions about the students.[7] By contrast, in deontic statements, which are used to report (rather than impose) a standing obligation or a permissible action, the surface order of the elements of the clause does not interfere with the content of the statement. The same holds in epistemically modalised utterances: passivisation does not affect epistemic evaluation (i.e. whether a certain proposition follows from/is compatible with the speaker's beliefs).

The same line of reasoning can explain another fact mentioned by Brennan (1993). She considers a range of symmetric predicates such as *shake hands with* and *get the same score as*, which license inference patterns of the form: R (x, y) ↔ R (y, x). In the following examples, whenever (56a) or (57a) is true, (56b) or (57b) respectively is also true (Brennan, 1993: 45-6):

(56) a. The governor shook hands with all the prisoners.
b. All the prisoners shook hands with the governor.

(57) a. Peter got the same score as Joan.
b. Joan got the same score as Peter.

It turns out that, if modal auxiliaries are inserted in these utterances, the inference patterns are disturbed under root interpretations but not under epistemic interpretations:

(58) a. The governor may shake hands with all the prisoners.
b. All the prisoners may shake hands with the governor.

(59) a. Peter can get the same score as Joan.

---

[7] The analysis can be extended to root readings beyond obligation and permission. For instance, several authors have considered the case of 'ability' *can* as particularly strong motivation for the claim that root modal readings impose selectional restrictions on the subject (Zubizarreta, 1982; Roberts, 1985):

(i) John can climb rocks.

On the account I am proposing, this fact about the subject of ability *can* is a direct consequence of the nature of the modal domain selected for the interpretation of (i). Recall that the modal domain for ability interpretations is typically subject-oriented: i.e. it includes assumptions from the file of the individual who figures in the sentential subject position. Therefore, in utterances such as (i) it matters what the subject is in order to derive the appropriate domain for interpreting *can*. No such restrictions are placed on the derivation of modal domain in simple root readings of *can* such as (ii):

(ii) Prices can vary.

b. Joan can get the same score as Peter.

On an epistemic reading of *may*, given (58a), it follows that (58b): if it is likely that the governor will shake hands with the prisoners, it follows that it is likely that the prisoners will shake hands with the governor. However, on a root reading of *may*, it no longer follows from the fact that the governor is allowed to shake hands with all the prisoners that the prisoners are allowed to shake hands with him. Similarly, root *can* blocks the inference from (59a) to (59b): while Peter may have the ability to get the same score as Joan (no matter what score she gets), Joan may not be able to get the same score as Peter. Substituting *can* with epistemic *may* renders the inference valid.

Rather than treating these facts as evidence for concatenating root modals with the VP in the semantics (cf. Brennan, 1993: 46), it is preferable to explain them in terms of the different accessibility of contextual assumptions in constructing the modal domains for each utterance. As before, the root pairs in (58) and (59) are not equivalent for pragmatic reasons: the surface order of the elements in each member of the pair provides a different starting point in constructing the relevant modal domain. For instance, although in a sense both members of (58) involve the 'same' permissible situation, (58a) views it from the point of view of the governor, while (58b) from the point of view of the prisoners. This is bolstered by the fact that the utterances, on a root reading, normally give rise to different implications. (58a) typically encourages inferences of the sort in (60), while (58b) typically invites inferences of the sort in (61):

(60) a. The governor won't be kept away from the prisoners.
b. The governor has to follow a schedule made for him by someone else.
c. The governor will not see everything in the prison.

(61) a. The prisoners are treated in a human way.
b. The prison authorities take good care of the prisoners.
c. There is no discrimination among the prisoners.

3.1.2.5 Properties of the Verbal Complement. Lastly, I wish to show that the characteristic properties of the complement, rather than being structurally determined by the type of main-clause modal (root or epistemic), correspond to pragmatic aspects of the interpretation process.

As before, the observations about how lexical aspect affects modal interpretation face a certain amount of counterexamples:

(62) a. You must have finished your homework by tomorrow.
b. One should have taken a language test before joining the course.

(63) a. One must be watching the children every minute, otherwise who knows what they'll come up with.
b. We must be leaving soon.

(64) a. I must be the best chess player there is.
b. You must believe in God, or they'll burn you at the stake.[8]

(65) a. You may go, but then again if you are not invited, you may end up not going.
b. They are good, they ought to break the encryption code.

(62)-(63) show that the presence of perfective or progressive aspect in the complement of the modal does not uniquely impose an epistemic interpretation. (64) demonstrates that individual-level stative verbs in the complement are not a sufficient condition for epistemic readings. Inversely, (65) shows that activity predicates in the complement do not require root readings.

One might object that the purported counterexamples are not well-designed to invalidate the claim that root interpretations broadly involve activities and epistemic interpretations (non-agentive) events. For instance, in (64a) *be (the best chess player)* is not actually interpreted as an individual-level stative predicate: rather, it is understood as marking the beginning of a state (one in which the speaker is the best chess player). This inchoative interpretation corresponds to an achievement, and achievements are activities. Likewise in (64b), *believe* is understood inchoatively as an achievement (i.e. the beginning of the state of believing); alternatively, it is understood as 'show that you believe/pretend to believe when asked'. On either of these interpretations, the complement of the modal denotes an activity, and therefore conforms to the aspectual prerequisites for root readings.

My response to this objection goes as follows. Recall that many authors view the clusters of aspectual properties as grammatical reflexes of root or epistemic meanings. On their view, the computation of the intended interpretation of modals presumably takes place in a completely structure-driven way. It is this strong view that the counterexamples in (62)-(65) were meant to question. The objection implies that these features might be seen as conceptual rather than grammatical requirements for root or epistemic modal

[8] Example from Steedman (1977).

interpretations. For instance, it gives good evidence that the requirement for a stage-level or activity predicate in the complement of root modals holds not on the level of semantic encoding (*be the best chess player* in (64a) and *believe* in (64b) are lexically individual-level statives) but on the level of conceptual structure. Consequently, the pragmatic reinterpretation of *be* and *believe* as activity predicates is imposed for conceptual reasons.[9] This conclusion is entirely in keeping with my own position, according to which the aspectual properties of modal interpretations listed in 3.1.1.6 correspond to natural aspects of the comprehension process. It is also compatible with the view that configurational properties of modals simply favour - rather than determine - either root or epistemic readings.

Let me turn to the future orientation of root modal complements. It is an inherent property of both granting permission and imposing an obligation that certain conditions obtain in the situation of utterance (cf. Searle, 1969). For instance, permission is typically granted for an activity (not a state), which will be carried out in the future (not the past or the present) by one or more responsible individuals; and likewise for obligation. These constraints are based on the strong assumption that it is not possible to influence past situations by acts carried out in the present due to the 'predominantly futureward direction of causation' (Lewis, 1973b: 36). Such independent considerations are responsible for the systematic priming of non-root interpretations of modals in complements marked by perfective or progressive aspect. More generally, they offer the conceptual backdrop for the association of root and epistemic modality with activities and (non-agentive) events respectively.

These conditions may be overriden in specific contexts. Assuming a fairy-tale context, which activates non-standard assumptions about causality, the utterance in (66) can be felicitously interpreted as granting permission:

(66) You may have been born in 1900.

Further support for the proposed analysis comes from the following fact: there are parallels in terms of aspectual profile between root (especially deontic) modal readings and a class of imperatives with the illocutionary force of permission or obligation. Such imperatives normally require activity or stage-level stative predicates and sound anomalous with individual-level stative predicates:

(67) a. Behave yourself.

---

[9] On the interaction of semantic and pragmatic factors in the computation of aspect, see Dowty (1986).

b. Be quiet.

(68) a. ?Be a native speaker of Japanese.
b. ?Know the answer.

Similarly, this class of imperatives has future orientation for reasons connected to illocutionary force:[10]

(69) a. Behave yourself at the party tonight.
b. ?Behave yourself at the party yesterday.

(70) a. Be quiet tonight.
b. ?Be quiet yesterday.

3.1.3 A Summary

I have surveyed several distributional facts about different readings of modal verbs, including scope with respect to negation, relative scope of root-epistemic interpretations, sentence-initial placement in interrogatives, selectional restrictions on the subject, and grammatical features of the complement. It appears that these facts cannot function as strict grammatical reflexes of root or epistemic modal readings in English; at best, they reflect pragmatic conditions on the sort of conceptual representation in which root or epistemic interpretations may felicitously occur.

It is worth pointing out that most syntactic analyses of modality have focused on epistemic and a subset of root (i.e. deontic) interpretations, thus ignoring other types of modality found in natural language. An exhaustive account would also have to include, for instance, alethic or simple root interpretations. It is unlikely, however, that every type in the full range of modal interpretations can be argued to differ from the rest on syntactic grounds. For instance, alethic and epistemic interpretations pattern uniformly with respect to both scope and the properties of the subject/complement of the complex; (71) can have either an alethic or an epistemic reading:

---

[10] Notice that imperatives with other illocutionary forces license different aspectual profiles. *Please know the answer* is felicitous as a wish (e.g. when one asks a passer-by for directions to the station). Likewise, wishes are not confined to future orientation: the passer-by example is interpretable as a wish about the present.

(71) I may win the jackpot.

I take such considerations to furnish additional support to the thesis that syntactic factors underdetermine, rather than streamline, the interpretation of modal expressions:

It might seem that a conceptual account of the distributional properties of deontic and epistemic modality fails to capture the intuition behind the raising-control alternation. In fact, a strong piece of evidence for the conceptual account comes from a recent discussion of raising and control phenomena by Ruwet (1991). According to Ruwet, the class of verbs (including modals) which exhibit mixed control-raising behaviour is far from homogeneous. After examining a wide array of data, Ruwet concludes that the alternation is not likely to be explained on syntactic grounds alone, and offers a tentative conceptual explanation based on what he calls an 'external vs. internal point of view' (Ruwet, 1991: 80). Although this is not the place to expand on his position, the idea that a purely formal approach to the raising-control alternation should be abandoned in favour of a conceptual account is very similar in spirit to the approach I have outlined in this section.[11]

Where does the present discussion leave us with respect to polysemy analyses of English modals? To the extent that distributional features of modal verbs do not uniquely determine the intended modal interpretation, it becomes implausible to view epistemic and root readings as two separate encoded clusters of modal meaning with different grammatical properties. Grammatical features, such as the aspectual profile of the complement, typically interact with contextual considerations to favour one or the other interpretation of a modal expression by contributing in specific ways to the main conceptual representation explicitly communicated by the utterance (i.e. the proposition expressed by the utterance).

3.2 Truth-Conditional Behaviour

3.2.1 The Standard View

It is widely assumed in the linguistic and philosophical literature that one of the crucial differences between root and epistemic modality is that the former, but not the latter, contributes to the truth-conditional content of the utterance. Specifically, it is generally

[11] This is not to say that our accounts do not differ; see Ruwet (1991: 81).

agreed that epistemic modality expresses some sort of comment on the proposition expressed by the rest of the utterance. Halliday (1970: 349), for instance, writes:

> [Epistemic modality] ... is the speaker's assessment of probability and predictability. It is external to the content, being a part of the attitude taken up by the speaker: his attitude, in this case, towards his own speech role as 'declarer'.

Lyons (1977) makes similar observations. Following Hare (1970), he distinguishes three levels of increasingly wide scope in the structure of utterances: the *phrastic*, which corresponds to the propositional content of the utterance; the *tropic*, which indicates the sort of speech act that the sentence is characteristically used to perform ('a sign of mood'); and the *neustic*, which indicates 'a sign of subscription' to the speech act performed - i.e. the speaker's commitment to the factuality, desirability, etc. of the base proposition (ibid. p. 749). Lyons places epistemic modality[12] outside the proposition expressed and within the neustic component of the utterance. In an utterance like (72), for instance, the speaker indicates that she entertains the embedded proposition 'There (will) be some rain tomorrow evening' with a weak degree of strength, and thus the hearer is recommended a correspondingly low degree of commitment to it:

(72) There may be some rain tomorrow evening.

More recently, Sweetser, in her analysis of the English modal system (Sweetser, 1990), suggests that epistemic modals do not contribute to what she calls the 'content' level of the utterance (cf. also Nuyts, 1993).

A number of authors have proposed that epistemic modality typically marks subjectivity in language, a property which is held to account for the non-truth-conditional status of epistemic modal items. Subjectivity is a rather loose term for the degree of the speaker's commitment and affective involvement in the content of the proposition she wishes to convey (Coates, 1983; Langacker, 1990; Sanders and Spooren, 1997). Palmer (1986: 54-5), for instance, suggests that epistemic modality in natural language is essentially subjective, in that it indicates 'the status of the proposition in terms of the speaker's commitment to it'. Likewise, Traugott (1982, 1988, 1989, 1995) views the rise of epistemic meanings of modal lexical items as an instance of a general historical development towards subjectification in language. After examining the meanings of various modal terms in

[12] In his terms, 'subjective epistemic modality'; cf. the discussion in section 2.4.3 above.

English (modal auxiliaries, speech-act verbs and modal adverbs) through gradual grammaticalisation from an initial root domain to the epistemic domain, Traugott concludes that meanings in natural language 'tend to be increasingly based in the speaker's subjective belief state/attitude toward the proposition' (Traugott, 1989: 35), and takes epistemic elements to be indicative of that tendency in that they 'express beliefs about the truth of the proposition' (ibid. p. 43).[13] Finally, Bybee and Fleischman (1995a) epitomise current views on epistemic modality by emphasising its relationship to the category of evidentiality and pointing out that 'epistemics are clausal-scope indicators of a speaker's commitment to the truth of a proposition' (Bybee and Fleischman, 1995a: 6).[14]

There are various types of evidence adduced to support the non-truth-conditional properties of epistemic modality. Lyons (1977: 799) mentions the resistance of epistemic modality markers to scoping under an attitude of doubt/rejection/acceptance or to attaching to the complement of a factive predicate or a verb of telling. The utterances in (74) are unacceptable as responses to the epistemic statement in (73), while those in (76) are fine given that the modal in the preceding utterance - (75) - is assigned a root reading:

(73) Alfred must be secretly seeing Barbara.

(74) a. ?Is that so? (= Is it the case that A. must be secretly seeing B.?)
b. ?I don't believe it. (= I don't believe that A. must be secretly seeing B.)
c. ?I agree. (= I agree that A. must be secretly seeing B.)
d. ?It is surprising that Alfred must be secretly seeing Barbara.
e. ?Mary told us that Alfred must be secretly seeing Barbara.

(75) The area must be evacuated.

(76) a. Is that so? (= Is it the case that the area must be evacuated?)
b. I don't believe it. (= I don't believe that the area must be evacuated)
c. I agree. (= I agree that the area must be evacuated)
d. It is surprising that the area must be evacuated.
e. Mary told us that the area must be evacuated.

---

[13] I return to Traugott's views on subjectification and the historical development of modal expressions in Appendix 3A and chapter 5.

[14] The last quote makes evident how the high scope of epistemics is linked to the argument concerning truth-conditionality.

Taken together with the point that epistemic modality indicators have clausal scope, such evidence is taken to suggest that epistemic modality in fact scopes over the proposition expressed by the (rest of the) utterance. The epistemic operator expresses the speaker's attitude towards this base proposition; consequently, it cannot co-occur with other attitude operators, as the unacceptability of (74a-d) illustrates. By contrast, (76a-d) are acceptable when understood as responses to the base proposition (i.e. the complement of the epistemic modal):

(77) a. Is that so? (= Is it the case that A. is secretly seeing B.?)
b. I don't believe it. (= I don't believe that A. is secretly seeing B.)
c. I agree. (= I agree that A. is secretly seeing B.)
d. It is surprising that Alfred is secretly seeing Barbara.

As a further test, consider placing utterances containing epistemic and root modals in the scope of logical operators:

(78) a. ?If John must have a high IQ, then his teachers should treat him carefully.
b. ?If that blonde may be Jack's wife, we should keep quiet about the secretary.

(79) a. If John must leave, then I will leave too.
b. If money may rule, then there's no justice.

As (78) illustrates, in English it is normally impossible to have epistemically used modals in the antecedents of conditionals. Root modals, by contrast, are subject to no such restriction, as shown in (79).

In conclusion, it appears that epistemic modal items are somehow sealed off from the elements of the proposition expressed by the (rest of the) utterance. In other words, it appears that epistemic modality markers introduce a multi-layered proposition, only part of which (:the proposition embedded in the modal) contributes to the truth conditions of the utterance. To illustrate, consider the following example of epistemic modality:

(80) 'Susanna Foo' must be one of the most expensive restaurants in Philadelphia.

According to the view outlined in the previous paragraphs, (80) communicates the double-level proposition in (80'):

(80') a. 'Susanna Foo' is one of the most expensive restaurants in Philadelphia.
b. As far as I know, it must be the case that (a).

(80'a) represents the proposition expressed by (80), or the truth-conditional content of (80). (80'b) is distinct from this basic (ground-floor), truth-conditional proposition; its role is to fine-tune the interpretation of the base proposition. Therefore, (80'b) is interpreted as a comment conveying the degree of speaker commitment to the proposition expressed by the utterance.

The behaviour of epistemic modality markers with respect to tests for truth-conditionality might seem to present a problem for the univocal analysis of modality advocated in chapter 2; independently of that, it is an interesting fact, which should be captured by any semantic/pragmatic analysis of modal phenomena. This is the way I propose to proceed: In section 3.2.2, I will discuss the idea that epistemic modal expressions (along with a variety of other constructions) introduce more than one act of communication, or a multi-layered proposition. In section 3.2.3, I will provide an alternative account which does not assume non-truth-conditional properties for epistemic modality. More specifically, I will show that my analysis of epistemicity in terms of metarepresentation can naturally accommodate the patterning of epistemic modal expressions with respect to the diagnostics for truth-conditionality.

For the moment, I want to consider some preliminary reasons for scepticism about the claim that epistemic modals do not contribute to the truth-conditional content of the utterance. Consider first the 'symmetry of enrichment' argument. I have argued so far that the variety of contextual interpretations of the English modals (simple root, deontic, etc.) is mostly the result of pragmatic enrichment of their underspecified semantic content. If epistemic interpretations are considered non-truth-conditional, the epistemic domain will be the only domain which does not contribute in the regular way to the construction of the proposition expressed. Although not decisive, this argument nevertheless provides some motivation for maintaining a symmetrical picture of the proposition expressed by utterances containing a modal verb (other things being equal).

One might retort here that the epistemic domain is, on anybody's account, different from other propositional domains. The standard view, as described above, holds that, to the extent that they involve the speaker's beliefs, epistemic interpretations are 'subjective', and are therefore not expected to contribute to truth conditions at all. However, the notion of subjectivity as standardly used in connection with epistemicity is singularly ill-equipped to help with the issue of truth conditions. The reason is that, as a number of authors emphasise, subjectivity is a graded notion, and modals differ with respect to their position on the continuum of subjectivity (Coates, 1983; Langacker, 1990; Palmer, 1986). The question then arises: what do degrees of subjectivity amount to in truth-conditional terms? For two-valued logic at least, gradience in subjectivity cannot be matched by a graded contribution to truth-conditional content. Hence, subjectivity does not settle the issue of truth conditions for epistemics; rather, it muddles it.

There is a second strong argument against the standard view, which appeals to our intuitions about truth conditions. Consider (81):

(81) a. The press may find this story interesting.
b. The press must find this story interesting.
c. The press will find this story interesting.

Suppose it turns out that the press does not find this particular story interesting. Does this make the utterances in (81a-b) on their epistemic interpretations false? I don't think so. What the speaker has said is that, as far as she knows, it is possible/necessary that the press will find the story interesting. (81a) and (81b) are clearly different from (81c), which would indeed be proven false in the circumstances I have described.

3.2.2 Multiple Acts of Communication

The idea that an utterance may communicate a multi-layered proposition is a familiar one. From the early speech-act literature onwards, a number of constructions or expressions have been thought to give rise to similar, double-level propositions. I will look at two cases which are frequently mentioned in connection with epistemic modality: evidential markers and parenthetical/performative verbs.

Consider first evidentiality. According to the standard (functional) definition, evidential markers show the kind of justification for a factual claim which is available to the person making that claim (Anderson, 1986; cf. Bach and Harnish, 1979; Urmson, 1963; Slobin and Aksu, 1982; Willett, 1988; Palmer, 1990; Mayer, 1990; Ifantidou, 1994). As a result, evidential expressions are generally expected to carry information about the source of the speaker's knowledge (observation; communication; inference; memory) and/or the degree of speaker certainty (strong/weak). Crucially, evidentials are not themselves the main predication of the clause, but are rather 'a specification added to a factual claim *about something else*' (Anderson, 1986: 274-5, his emphasis).

Consider the examples of evidential and hearsay adverbials in (82) and (83)[15]:

(82) Obviously, Mark is going to get the job.

(83) Allegedly, Mark is going to get the job.

The utterances in (82) and (83) are assumed to give rise to the double-level propositions in (82') and (83') respectively:

(82') a. Mark is going to get the job.
b. It is obvious that Mark is going to get the job.

(83') a. Mark is going to get the job.
b. It is alleged that Mark is going to get the job.

According to standard speech-act analyses (Bach and Harnish, 1979; Urmson, 1963), the propositions expressed by (82) and (83) are (82'a) and (83'a) respectively. The propositions (82'b) and (83'b) are also communicated by the original utterances, although their role is a different one: 'They help the understanding and assessment of what is said rather than being part of what is said' (Urmson, 1963: 240). In (82), the sentence adverbial *obviously* marks the evidential support which the ground-level proposition (82'a) receives in the speaker's cognitive environment. In (83), the adverbial *allegedly* marks the source of the information conveyed by the base proposition as external to the current speaker (i.e. it marks the information as a product of communication).

The second class of items I want to examine here includes parenthetical verbs such as *I believe, I conclude, I guess*, etc. (and, secondarily, certain performatives such as *I admit, I infer, I presume,* etc.). The function of these items is generally seen as very similar to that of the parenthetical adverbials introduced in the previous paragraphs. Urmson (ibid. p. 225) summarises the role of parenthetical verbs in the following way:

> We make our statements in contexts, social as well as logical. For example, we often have an emotional attitude to the fact we state... Further, we make our statements sometimes with good, sometimes with moderate, sometimes with poor evidence... It is my contention that parenthetical verbs are some of the devices that

---

[15] Note that, on this definition, hearsay markers are a subtype of evidentials.

we use to deal with these matters... By them we prime the hearer to see the emotional significance, the logical relevance, and the reliability of our statements.

To illustrate, consider the pair in (84)-(85):

(84) I admit that I was unaware of the crisis. (said by the President)

(85) The President was, I believe, unaware of the crisis.

The above utterances are seen as communicating the multi-layered propositions in (84') and (85'):

(84') a. The President was unaware of the crisis.
b. The President admits that he was unaware of the crisis.

(85') a. The President was unaware of the crisis.
b. The speaker believes that the President was unaware of the crisis.

As before, the truth-conditional content of the utterances in (84) and (85) is assumed to be given by (84'a) and (85'a). The propositions in (84'b) and (85'b) are considered to be indications of how the base propositions are to be taken. In the first case, *I admit* indicates how the base proposition 'fits logically into the discourse' - 'one is not reporting the occurrence of a bit of admitting, whatever that may be supposed to be' (Urmson, 1963: 226). In the second case, *I believe* marks the degree of reliability of the ground-floor proposition (cf. Bach and Harnish, 1979).[16]

It is easy to see why epistemic modality has been traditionally associated with evidentiality and parenthetical verbs such as *I conclude, I guess*, etc. Modal verbs, on their epistemic interpretations, convey information about both the source of the speaker's knowledge (i.e. inference) and the degree of speaker certainty (strong in *must*, weaker in *may*): in this sense, they can be said to properly belong to the class of evidentials.[17]

[16] Grice (1989) has made similar suggestions about sentence connectives such as *but, therefore, moreover*, and parenthetical expressions such as *on the other hand*: what the connectives or parentheticals contribute to communicated meaning is an indication of a higher-order speech act which comments in a certain way on the ground-floor proposition. For a relevance-theoretic analysis of these expressions, see Blakemore (1987).

[17] Here, as above, I adopt a functional characterisation of evidentiality. There are alternative, semantic definitions of evidentiality in the literature: according to Anderson (1986), genuine evidentials are those expressions which primarily *encode* information about evidence. It follows that, on a unitary semantic

Likewise, epistemic modals mark the proposition embedded under them as a conclusion with variable degrees of support from available evidence: in this sense, they may be compared to parentheticals such as *I infer, I conclude*, and so on. In all these cases, the modals seem to indicate how the hearer is to 'take' the proposition expressed by the rest of the utterance.

Still, I want to argue that the parallel with sentence adverbials and parenthetical verbs is not quite accurate. The reason is that it overlooks the crucial fact that epistemic modals are not parenthetical expressions: while evidential and hearsay adverbials and parenthetical verbs are only loosely attached to the rest of the utterance, both syntactically and phonologically,[18] epistemic modals are the main predication of the utterance. Whereas parenthetical expressions may be seen as contributing to non-basic propositions which simply fine-tune the interpretation of the ground-floor, truth-conditional representation communicated by the utterance, epistemic modals cannot easily be detached from the ground-floor proposition (e.g. through comma intonation, or free word movement). This fact squares well with the observation made earlier that we do have intuitions which treat epistemic modality as part of the truth-conditional content of the utterance. It seems, therefore, that epistemic modality cannot be handled in exactly the same way as evidential sentence adverbials and parentheticals, and that the parallel between the two classes of constructions has to be reformulated. Incidentally, epistemic modals seem to behave in this respect more like the class of performative verbs such as *I infer*, *I conclude*, etc. in their non-parenthetical uses: these verbs are also syntactically fully integrated, and form a non-detachable part of the utterance.

To prepare the ground for a more detailed comparison, I would like to look at some recent analyses of examples of the type in (82)-(85) by Elly Ifantidou and Diane Blakemore within the relevance-theoretic framework. Ifantidou (1994) provides a detailed analysis of evidential and hearsay adverbials as well as parenthetical verbs. Her analysis generally preserves the assumption that both parenthetical adverbials and parenthetical verbs do not contribute to the proposition expressed by the utterance; their role is to participate in the construction of explicitly communicated higher-level propositions (higher-level explicatures) - cf. the propositions (82'b), (83'b), (85'b) above.[19] According

---

account of modals, epistemic interpretations would fail Anderson's semantic criterion of evidentiality (since they would be computed through pragmatic inference and not semantic decoding).

[18] For a thorough treatment of such disjunct constituents, see Espinal (1991).

[19] In fact, Ifantidou recognises truth-conditional as well as non-truth-conditional uses of evidential and hearsay adverbials; in this respect, her account differs from speech-act accounts which insist on the non-truth-conditionality of all sentence adverbials (including illocutionary and attitudinal adverbials). For instance, she claims that truth-conditional uses can fall under the scope of logical connectives, as in (i) and (ii) below:

to Ifantidou, the role of these higher-level propositions in the above examples is to fine-tune the comprehension of the base proposition (i.e. the proposition expressed by the utterance) in specific ways. For instance, evidential adverbials indirectly affect the strength with which the base proposition is held. The degree of strength communicated obviously varies according to the semantic content of the evidential adverbial. Weak evidentials (*apparently, seemingly*, etc.) indicate a reduced degree of strength; strong evidentials (*obviously, clearly*, etc.) indicate a high degree of strength. By affecting the degree of strength attached to the base proposition, both weak and strong evidentials also affect the degree of strength recommended to the hearer.

Hearsay adverbials are a different case. Their role is to mark the base proposition as a case of interpretive use. (83'b) above, for example, attributes the base proposition to someone other than the current speaker, thereby waiving the present speaker's direct commitment to it. Other examples may communicate an indirect commitment to the base proposition, in case the current speaker trusts its original source.

Finally, Blakemore (1990/1) has analysed the class of performatives such as *I predict, I infer, I conclude*, and so on. Following Sperber and Wilson (1984), she takes these to correspond to *non-communicated* speech acts (and calls them *non-communicated performatives*). Making a prediction, for instance, does not involve the formation and recognition of manifest intentions about speech-act type; the necessary conditions for predictions have to do with the content communicated - hence prediction is not a communicated act.[20] The gist of Blakemore's account is that non-communicated performatives give rise to a double-level proposition. She argues against speech-act accounts and claims (with Recanati, 1987) that such performative verbs should not be seen as external to the proposition expressed by the utterance. Although an utterance such as (84) cannot be seen as describing an existing state of affairs or reporting one of the speaker's beliefs, it does represent a state of affairs. For instance, in producing (84) the

---

(i) a. The cook obviously won't poison the soup.
   b. If the cook obviously won't poison the soup, we can eat the meal without worrying.
(ii) a. The cook has allegedly poisoned the soup.
   b. If the cook has allegedly poisoned the soup, the police should make an inquiry.

However, we still do not have adequate justification for the presence and distribution of both truth-conditional and non-truth-conditional uses for evidential and hearsay adverbials. Ifantidou simply notes that 'a parenthetical comment which... merely expresses the speaker's attitude to the fact that [the ground-floor proposition] is true... will be perceived as non-truth-conditional' (1994: 189).

[20] Naturally, if there is some doubt as to how the hearer is to process the utterance, a speaker aiming at optimal relevance could stipulate explicitly that a prediction is being made - for instance, by using a 'non-communicated' performative such as *I predict, I suggest, I hypothesise, I guess, I conclude,* etc. Since predicting, suggesting, etc., do not rely for their use and comprehension on institutional facts (unlike baptising, blessing, etc.), they belong to *non-institutional* non-communicated acts (Sperber and Wilson, 1984: 10ff.).

speaker may be presenting evidence for the truth of the proposition in (84'b). Blakemore therefore concludes that an utterance such as (84) is used by the speaker to perform two distinct acts of communication. On the one hand, the speaker is communicating the information that he was unaware of the crisis (base proposition), while on the other hand, he is admitting that the base proposition is true. Crucially, Blakemore claims that the main relevance of an utterance such as (84) is achieved on the level of the embedded proposition: it is the information that the President was unaware of the crisis that will cause the main bulk of cognitive effects. The second act of communication (i.e. the act of admitting) achieves its relevance by helping the hearer process the first act. Therefore, the performative verb functions as a kind of conceptual constraint on the way the relevance of the embedded proposition is to be computed.

We now have the machinery with which to reconsider the role of epistemic modality in communication and to reformulate the parallel with the two classes of expressions above. I want to argue that epistemic modals bear some similarities to the class of evidentials and parentheticals in that both classes introduce metacognitive operations. Furthermore, I want to propose that epistemic modal verbs resemble non-communicated performatives in that both classes contribute to the truth-conditional content of the utterance, even though they give rise to a layered proposition. The following section is devoted to substantiating these claims.

3.2.3 Metarepresentation and Truth Conditions

Recall example (80), repeated below as (86):

(86) 'Susanna Foo' must be one of the most expensive restaurants in Philadelphia.

I suggest that the proposition expressed by (86) has the complex form in (86'):

(86') $[_p$['Susanna Foo' is one of the most expensive restaurants in Philadelphia]] is entailed by D_{bel}.

I have introduced double square brackets in (86') to highlight the fact that the embedded proposition p is metarepresentational, since at this stage in the discussion this property of p is going to be critical. Consider now how the hearer will process (86'). It is possible that this complex proposition will be capable of causing some cognitive effects as it stands.

Imagine, however, that (86') fails to interact with existing contextual assumptions to yield adequate cognitive effects. The hearer is then entitled to temporarily detach p from the metarepresentational environment in which it appears, and process it on its own. The proposition in (86') will consequently be analysed into the two components in (87):

(87) a. 'Susanna Foo' is one of the most expensive restaurants in Philadelphia.
b. (a) is entailed by the set of the speaker's beliefs.

The hearer can now arrive at a range of cognitive effects by combining (87a) with other assumptions available to him: he might draw some conclusions as to the type of restaurants, their clientele, whether he might be able to eat there himself, and so on. The role of (87b) will be to adjust the level of support each of these conclusions finally receives by indicating the strength of one of their premises (the proposition in (87a)).

Such temporary disquotations of metarepresentational material seem to be quite regular thought processes: they are formal procedures which operate on well-formed propositional forms. Some further examples are given in (88) (see Sperber, 1997, for discussion):

(88) a. Fodor believes that simple concepts are innate.
b. They say that it's better to be safe than sorry.
c. I hear that he is still abroad.

To take just one case: after detaching the embedded proposition in (88a) and processing it together with the assumption in (89a), the hearer may re-embed the product of this process under the initial metarepresentational environment and produce (89b):

(89) a. ELECTRON is a simple concept.
b. Fodor believes that ELECTRON is innate.

Going back to modality, it appears that epistemic modal verbs do contribute to the truth-conditional content of the utterance. However, it often turns out that the complex proposition to which they give rise cannot achieve relevance as it stands. The hearer, therefore, has to temporarily detach the embedded proposition and process it on its own in order to derive the main bulk of cognitive effects which the utterance was intended to create. This is done in accordance with the much more general process of disquoting propositions from metarepresentational environments in the language of thought and processing them independently in a temporary inferential space. After a satisfactory range of cognitive effects has been reached, the contribution of the epistemic modal verb is to modulate these effects according to the degree of support which the speaker's belief-set assigns to the proposition originally embedded under the modal.

On this picture, the truth-conditional import of epistemic modals is kept separate from their contribution to the (main) relevance of the utterance. Mixed intuitions about the role of epistemic modals can now be explained as a result of confusion between these two aspects. On the one hand, as I have already mentioned, one should in principle try to give an analysis of epistemic modals that is symmetrical to that of root modals (unless there are strong reasons to the contrary); the above solution indeed preserves the 'symmetry of enrichment' intuition. Furthermore, it is often felt that the content of an epistemically modalised utterance is not falsified by an outcome which falsifies the proposition embedded under the modal. This is now explained by the fact that the speaker in an utterance such as (86) is simply committed to the truth/factuality of the complex proposition in (86') - and not the truth/factuality of the embedded proposition p. A genuine falsification of the proposition expressed by (86) would be (among other things) the discovery of a counterexample to p which the speaker manifestly knows but which she has overlooked in evaluating p. On the other hand, the proposed account of epistemic modality captures the intuition underlying previous analyses that epistemic modal verbs make a comment on a (base) proposition, or an indication of speaker commitment. However, this is taken to be not an intuition about truth conditions, but about the function of epistemic modals in establishing the relevance of the overall utterance.

It is clear that epistemic modal verbs, on the disquotational account I propose here, bear various similarities to the class of performative verbs analysed by Blakemore (1990/1). Performative verbs such as *I conclude, I infer, I predict*, and so on, in their syntactically non-parenthetical uses, regularly contribute to the truth-conditional content of the utterance. However, they do not participate in the computation of the main relevance of the utterance but are used to regulate in a specific way the bulk of cognitive effects which the main (base) proposition achieves. In (90), for instance, the contribution of the main verbs to the relevance of the utterance depends on the prior computation of the range of effects which the proposition in (91) yields in each case:[21]

[21] One could speculate at this point about what triggers the layered proposition effect in the case of non-communicated performatives. It could be that verbs such as *I conclude, I suggest*, etc. are also metarepresentational environments, so that the double-level analysis marks the disquotation of the embedded representations. Another consideration is the syntactic partitioning of the utterances in (40) into main (performative) verb and complement. Blakemore's suggestion about multiple acts of communication takes a somehow different perspective on these issues.

(90) a. I conclude that the Earth is flat.
b. I infer that the Earth is flat.
c. I insist that the Earth is flat.
d. I suggest that the Earth is flat.

(91) The Earth is flat.

So far, so good. However, we still lack an explanation of the fact that epistemic modals do not satisfy the normal tests for truth conditionality. Next I want to claim that, far from being surprising, the behaviour of epistemic modals with respect to these tests is, in fact, to be expected. As a reminder, I repeat below the examples in (73) - (76), which were taken as *prima facie* evidence for the non-truth-conditionality of epistemic modal verbs:

(92) Alfred must be secretly seeing Barbara.

(93) a. ?Is that so? (= Is it the case that A. must be secretly seeing B.?)
b. ?I don't believe it. (= I don't believe that A. must be secretly seeing B.)
c. ?I agree. (= I agree that A. must be secretly seeing B.)
d. ?It is surprising that Alfred must be secretly seeing Barbara.
e. ?Mary told us that Alfred must be secretly seeing Barbara.

(94) The area must be evacuated.

(95) a. Is that so? (= Is it the case that the area must be evacuated?)
b. I don't believe it. (= I don't believe that the area must be evacuated)
c. I agree. (= I agree that the area must be evacuated)
d. It is surprising that the area must be evacuated.
e. Mary told us that the area must be evacuated.

It seems to me that the impossibility of getting responses such as (93a-d) to the epistemic (92) can be straightforwardly explained within the framework I have developed. It is reasonable to assume that, in each of these responses, the speaker is reacting to the main point made by the previous utterance. In accordance with what was claimed earlier, the main point of (92) is the embedded proposition 'Alfred is secretly seeing Barbara' - therefore, this is what the speaker's doubt, disbelief, agreement or surprise picks out. In the root example, by contrast, the modal verb forms part of a single proposition and therefore makes a contribution to the main relevance of the utterance. As a result, the responses in (95a-d) are acceptable.

(93e) is more complicated. Recall that the complex proposition expressed by an epistemically modalised utterance makes reference to the speaker's belief-set (see, e.g., clause (87b) above). This introduces an element of indexicality into the representation: modal verbs like *must, may, should*, and so on, on their epistemic interpretations, involve the evaluation of a proposition with respect to the current belief-set of the speaker in the here-and-now of the talk-exchange. This property of epistemic modality is responsible for the unacceptability of (93e). Barring verbatim reported speech and full mention in general, (93e) is ill-formed because it is impossible to use *must* to refer to Mary's original epistemic evaluation of the embedded proposition - the verb will default to the current speaker and her present evaluation of whether Alfred is secretly seeing Barbara. No such property inheres in the root interpretation of (94), so that the response in (95e) is perfectly well-formed.

The indexicality of epistemic modals is also responsible for their inability to fall under the scope of conditionals. Recall the previous examples:

(96) a. ?If John must have a high IQ, then his teachers should treat him carefully.
b. ?If that blonde may be Jack's wife, we should keep quiet about the secretary.

(97) a. If John must leave, then I will leave too.
b. If money may rule, then there's no justice.

In (96), *must* and *may* are used for the on-line performance of an inference on the part of the speaker. As such, their content cannot be placed in the antecedent of a conditional. Take (96a): it is difficult to see what conclusion could follow from the fact (roughly) that the speaker's beliefs entail that John has a high IQ. On a possible reformulation of (96a), after omission of the modal, only the content of the embedded proposition falls under the scope of the conditional - and the result is a well-formed utterance. Root modals can unproblematically contribute their conceptual content to an *if*-clause, as (97) shows.

On this analysis, then, the alleged diagnostics for truth conditionality do not in this case, in fact, test for contribution to truth conditions. At most what they pick out is the main point of the utterance. A further argument to this effect is that non-communicated performatives, which do contribute to the truth-conditional content of the utterance, also resist falling within the scope of the conditional. Notice that (by definition) performatives are tied to the here-and-now of the utterance and, rather than describing a certain state of affairs, they mostly contribute to relevance by fine-tuning the interpretation of the proposition embedded under them:

(98) a. ?If I conclude that the Earth is flat, then I shouldn't like to go beyond San Francisco in case I fall off.
b. ?If I insist that the Earth is flat, then I shouldn't like to go beyond San Francisco in case I fall off.

If the parallel between epistemic modals and a certain type of performatives is close enough, some other parallels drawn in previous analyses of modality will have to be reconsidered. I have in mind primarily the connection with various parenthetical expressions, notably verbs and adverbials. On the analysis I have presented here, there are syntactic reasons for keeping the class of parentheticals distinct from that of modal verbs. Parenthetical elements, being only loosely attached to the base clause, cannot be taken to contribute to the (main) proposition expressed by the utterance. Their role is to contribute to higher-level explicatures which fine-tune the comprehension of the main propositional content of the utterance. Consider the following examples:

(99) a. According to Fodor, simple concepts are innate.
b. As people say, it's better to be safe than sorry.
c. He is still abroad, I hear.

(100) a. The Earth, I conclude, is flat.
b. The Earth, I infer, is flat.
c. The Earth, I insist, is flat.
d. The Earth, I suggest, is flat.

(99) is a parenthetical version of (88) and (100) a parenthetical version of (90). By way of illustration, I give the proposition expressed by (99a) and (100a) together with the higher-level explicatures which the utterances communicate:

(101) a. Simple concepts are innate.
b. According to Fodor, (a).

(102) a. The Earth is flat.
b. The speaker concludes that (a).

As (101b) and (102b) are meant to show, higher-level explicatures are themselves metarepresentations. The difference from the non-parenthetical versions of these examples is that, in (99) and (100), the presence of a double-level utterance is made structurally transparent: two claims are being made by each member of (99) or (100), each with its own separate truth conditions, and only one of them contributes to the main truth-conditional content of the utterance. By contrast, in epistemic modals (and, arguably, non-parenthetical performatives), the speaker is committed to the truth of a complex claim, which is subsequently analysed into a double-level proposition (the proposition expressed by the utterance).

The distinction carries over to evidential and hearsay adverbials. Without going into much detail, I propose to generalise to the broader class of evidential markers the idea - already put forth for hearsay items - that these markers introduce interpretively used propositions. If evidentiality involves an appreciation of the validity and source of information, it should rest on the human ability to handle mental representations of external reality *qua representations,* and to focus on properties of their representing dimension. It follows that all of the arguments I have adduced for treating epistemic modality as metarepresentational apply automatically to the broader category of evidentiality (see also chapter 4).

Still, evidential adverbials such as *obviously, clearly, apparently*, etc., and hearsay adverbials such as *allegedly, reportedly*, etc., make a very different contribution to what is conveyed by an utterance from epistemic modals. Extending proposals by Ifantidou (1994),[22] we can view the two classes of adverbials as setting up a metarepresentation (captured by a higher-level explicature in the relevance-theoretic framework): the type of metarepresentational environment (i.e. the specific type of adverbial) will fine-tune the interpretation of the ground-floor proposition (i.e. the proposition expressed by the utterance). Hence, even though both epistemic modals and evidential adverbials set up metarepresentations, they differ in that only the first contribute their content to the truth-conditional representation of the utterances containing them.

The point is a subtle one. Interestingly, it seems that my position on epistemic modals affords a number of correct predictions about the truth-conditional behaviour of modal expressions in general. Firstly, alethic or logical interpretations of modality, although they involve metarepresentation, should not cause truth-conditional difficulties, since they are not indexical in the sense described above for epistemic modals. This prediction is confirmed by examples such as the following:

---

[22] See also Blakemore (1994). Her account does not apply the concept of metarepresentation to evidentials (hearsay items excluded).

(103) If the sum of all angles in a triangle must be 180°, then my answer to the geometry exam is all wrong.

Secondly, modal adverbs such as *possibly, probably, necessarily*, etc., on their epistemic interpretations, should not contribute to the proposition expressed by the utterance. This seems to be right:

(104) If the guests possibly want to drink more, there is extra stock of wine in the cellar.

Modal adjectives behave just like modal verbs in this respect. (105), epistemically interpreted, expresses a complex proposition of the sort in (105'), much as (86) above did (for further data, see Nuyts, 1993):

(105) It is possible that Arsenal will win the cup.

(105') $[_p$[Arsenal will win the cup]] is compatible with D_{bel}.

Finally, alleged counterexamples to the proposed account can be reanalysed away quite straightforwardly. For instance, certain so-called epistemic uses of modal expressions which appear in environments where genuine epistemic uses are usually unacceptable, can and should be reinterpreted as root uses:

(106) If Paul may get drunk, I'm not coming to the party.

(107) A: Paul will be the one who gets a promotion.

(108) B: Not necessarily: Judy has been working very hard lately.

In (106), the speaker is taken to communicate that she is not coming to the party on the condition that there is an objective possibility that Paul will get drunk. This seems to suggest that we are, in effect, dealing here with a case of root modality. Indeed, in (106) the speaker does not make her coming to the party conditional on an epistemic relation, but on a factual, real-world possibility, supported by the circumstances in the world. The exchange in (107)-(108) is a second apparent counterexample to the view that epistemic modal adverbs do not contribute to the proposition expressed by the utterance and hence cannot scope under negation: nevertheless, I think *necessarily* should be given a logical, not an epistemic, reading here. What the speaker communicates by (108) is that there is another, logical rather than epistemic, possibility which the previous interlocutor has overlooked. Since the main point of my argument concerns English modal verbs, I will not go into the properties of other modal expressions in any greater detail.

In conclusion, let me consider the implications of the discussion for the polysemy-monosemy debate. A standard argument for the claim that the root-epistemic distinction was encoded in the English modal system was that the distinction coincided with truth-conditional vs. non-truth-conditional aspects of communicated content. Semantic accounts based on a unitary analysis of root and epistemic modals had to face the objection that it was only root interpretations that regularly contributed to the proposition expressed by the utterance. The discussion in this section has offered some arguments for treating both types of modal interpretations as part of the basic truth-conditional content communicated by the utterance. Moreover, the tests which claimed to prove that epistemic modality was external to the proposition expressed were reanalysed mainly as diagnostics which pick out that part of the utterance which is responsible for the main bulk of cognitive effects (or the main relevance of the utterance). Consequently, the argument from truth conditions does not seem to lend any support to a polysemic organisation of the semantics of the modals.

3.3 'Speech-Act' Modality

3.3.1 Previous Proposals

I now want to turn to a different area which might be thought to pose a challenge to semantic and pragmatic accounts of modality: the category of 'speech-act modality'. The term belongs to Eve Sweetser (1990), who introduces the new category as an extension of her basic polysemy-based account of English modals.

Recall that, on Sweetser's analysis, modals encode force-dynamic concepts which are metaphorically projected from the concrete, external world of socio-physical experience to the abstract, internal world of reasoning and of mental processes in general. Force-dynamic notions applied to the external world give rise to root modals, and epistemic modals are the counterparts of these concepts in the abstract world of mental processes. The category of speech-act modality is then introduced as a further mapping of force dynamics, this time into the domain of speech acts, as the following examples are meant to illustrate:

(109) He may be a university professor, but he sure is dumb.

(110) There may be a six-pack in the fridge, but we have work to do.

Sweetser observes that what the speaker wants to convey by (109) is something like 'I admit that he is a university professor, and I nonetheless insist that he is dumb'. Similarly, in (110), the speaker is responding to an offer previously made by her interlocutor by communicating something like 'I acknowledge your offer, and I nonetheless refuse it'. These readings are contrasted with the interpretations of (111), where *may* carries normal epistemic meaning (Sweetser, 1990: 70):

(111) a. He may be a university professor, but I doubt it because he is so dumb.
b. There may be a six-pack in the fridge, but I'm not sure because Joe had friends over last night.

Sweetser concludes that modality in (109) and (110) is applied not to the content world or the epistemic world but to the conversational world: 'the interlocutor is being *allowed* by the speaker to treat a certain statement as appropriate or reasonable, or to present an offer' (ibid. p. 71). Therefore, she says, the two utterances can be paraphrased as follows:

(109') I do not bar from our (joint) conversational world the statement that he is a university professor, but...

(110') I do not bar from our conversational world your offer of beer, but...

So speech-act modality is the result of applying the modal concepts to the conversational interaction itself: 'the speaker (or people in general) is forced to, or (not) barred from, *saying* what the sentence says' (ibid. p. 73). Although it is more difficult to find examples parallel to (109) and (110) using other modals, Sweetser mentions some more cases which seem to involve all of the modals quite regularly:

(112) Editor to journalist:
OK, Peking can be Beijing: but you can't use 'Praha' for Prague.

(113) Mondale advisor giving directions to speech writer:

Reagan must be a nice guy (as far as the content of the speech is concerned), even if we criticise his policies.

(114) To smoker of long cigarette, from speaker who recognises that 'cigar' dialectally signifies 'long cigarette':
In New Orleans, you would be smoking a cigar right now.

In justifying the introduction of speech-act modality into a model which so far includes root and epistemic meanings, Sweetser notes that 'an utterance is content, epistemic object and speech act all at once' (ibid. p. 75). Therefore, linguistic meaning may be grounded in any of the three conceptual domains of real-world (:content), epistemic and speech-act objects. Very often, a single meaning starts out from the more basic, real-world domain and ends up having counterparts in the other two domains. The same sort of polysemic structure is detected by Sweetser in conjunctions, conditionals, and other areas of natural language (see also Fauconnier, 1985, 1997).

Sweetser recognises that her account of speech-act modality needs further refinement, and acknowledges a number of puzzling points. First, (109) and (110) do seem different from the other examples of speech-act modality. Moreover, *may* cannot be used freely in utterances like (112)-(114): for instance, it gives strange results if substituted for *can* in (112). The explanation Sweetser puts forward is that *may* has been specialised for the sort of use exemplified in (112)-(114) - a proposal very similar to that found in Fillmore, Kay and O'Connor (1988), who view utterances such as (109) and (110) as cases of *constructions*: 'clusters of information including, simultaneously, morphosyntactic patterns, semantic interpretation principles to which these are dedicated, and, in many cases, specific pragmatic functions in whose service they exist' (ibid. p. 534). In a more recent discussion of examples like (109) within the framework of Construction Grammar, Paul Kay has maintained that the construction view can explain why, in (109), it is not *affirmed* that he *may* be a university professor (as it would if (109) were an epistemic statement), but rather it is *conceded* that he *is* a university professor. We could know everything else we know about the grammar and meaning of the words in (109), Kay argues, without knowing that the utterance means something different from, say, an epistemic statement (Kay, 1997: 51).

Both Sweetser's and Kay's approaches, then, have to attribute to the *may*-structures in (109) and (110) some degree of idiomaticity, in the sense that their interpretation cannot be predicted solely from the semantics of their separate parts.[23] An account which could show that the interpretation of the two examples follows quite naturally from their compositional semantics plus general pragmatic considerations would avoid the appeal to idiomaticity. If usual considerations of cognitive economy hold, such an account (if it could be made plausible) might be preferable on methodological grounds. One of my claims later on will be precisely that no construction-specific knowledge is necessary in order to derive the concessive overtones of these utterances.

Another point made by Sweetser casts further doubt on the hypothesis that there is a separate, homogeneous category of speech-act modality: cases like (113) and (114) seem to involve a use/mention distinction, 'in that the speaker is applying the relevant modality to the choice of linguistic *form*, not to the content' (Sweetser, 1990: 72).[24] However, Sweetser says, in utterances like (113) it is the *purport* of the speech which is in question, and so speech-act modality applies to the choice of a given content. One may wonder whether there could be a single, more explanatory category which would encompass the uses in (109)-(114) and naturally allow for either content or form to fall under the scope of modality.

The previous point raises a more general question as to the exact status of the speech-act domain with respect to the semantics/pragmatics distinction. Sweetser clearly intends the content and epistemic domains to have a bearing on linguistic semantics. Indeed, she says that some of the English modals actually encode both root and epistemic concepts. It is less clear that she should wish to extend this line to the speech-act domain. It certainly does not make much sense to claim that, say, *may encodes* speech-act modality, and Sweetser herself carefully avoids any such formulation. At best, then, one can speak of speech-act *uses* (rather than senses) of modal expressions. A pragmatic extension of modality into the speech-act domain also seems to fit the parallel drawn by Sweetser between speech-act modality and metalinguistic negation (Horn, 1985):

(115) I'm not his daughter; he's my father.

I will have more to say on the parallel in section 3.3.4. For the moment, I merely intend to point out that, even on Sweetser's analysis, speech-act modality has to be set apart from

---

[23] This is not to say that there are no differences between Kay's and Sweetser's analyses (see Kay, 1997: 51, fn.3).

[24] In a footnote, Sweetser leaves room for a separate subtype of speech-act modality to cover applications of modality to the form of the utterance (1990: 155); this would presumably cover examples (112)-(114), which she calls 'metalinguistic'. If my reanalysis of the speech-act examples below is right, these uses are explicable in a natural way.

the root and epistemic cases in that it cannot be viewed as systematically generating polysemy in natural language.

There is one more argument to this effect. Sweetser suggests that her tripartite modal structure may be semantically analysed in terms of Fauconnier's theory of mental spaces (Fauconnier, 1985): the content, epistemic and speech-act domains correspond to mental spaces, and a single root modal expression can have a counterpart in one of the other two spaces. Fauconnier's original conception of spaces was meant to capture referential 'ambiguities', which essentially stemmed from various ways of conceptualising a real-world referent. For instance, separate spaces were set up to deal with various mental representations of real-world referents (e.g. referents in pictures, or referents enmeshed in counterfactual situations). It is, therefore, quite easy to fit content and epistemic modality into this model: content modality would operate in the basic mental space which is used to represent real world entities and states of affairs, while epistemic modality would operate in a derived space which includes our mental representations of those entities or states of affairs (indeed Fauconnier discusses a number of epistemic expressions such as *in X's view*, *X believes that*, *according to X*, etc., which he considers space-builders).

However, mental space terminology does not apply as straightforwardly to speech-act modality. The question is: what would a mental space look like in this case? Most probably it would have to be a domain including descriptions of speech acts; the claim would then be that we conceptualise utterances as speech acts using a discrete speech-act mental space. However, it is not clear that the basic modal concepts of possibility and necessity would coherently apply to such a space. Consider example (109): what the speaker conveys is not that a *speech act* is possible or admissible, but rather that the *content* of a statement is possible or admissible with respect to certain background assumptions. The same holds for the other utterances in (110)-(114). Note that Sweetser (1990: 74) considers certainty as the counterpart in the epistemic domain of compulsion in the real-world domain, and epistemic possibility as the counterpart of root possibility or permission: however, she does not give any example of counterparts in the speech-act domain.

Finally, as Sweetser herself points out (ibid. p. 73), speech-act uses of the modals have a relatively restricted range of interpretations. For instance, they generally do not refer to the speech act being performed by the speaker. Consequently, one could not use (116) to communicate (117):

(116) Your father must want you home.

(117) I must tell you that your father wants you home.

Moreover, only certain types of modality can be felicitously expressed towards the speech act being performed. Although it is not rare to request permission, it is rare to assert it:

(118) May I ask you where you are going?

(119) ?I may ask you where you are going.

In the next section, I will propose an alternative approach to Sweetser's original examples. My main argument will be that speech-act modality is not a separate and natural class of phenomena but can be given a more plausible analysis in terms of concepts already independently motivated within relevance theory.

3.3.2 A Reanalysis in Terms of Metarepresentation

Let me go back to the utterances in (109) and (110). A closer look reveals that they share a number of features:

a) They occur most naturally as rejoinders to a previous utterance, some aspect of which was disputed by the speaker but is now grudgingly conceded. For instance, (109) might be a follow-up to the clause in (120c) in an argument about Jones' abilities:

(120) a. X: I admire Jones immensely. I think he has contributed a lot to the intellectual life of this country.
b. Y: I don't agree at all. In fact, I think the guy is completely incompetent.
c. X: Look, he is a university professor.
d. Y: He may be a university professor, but he sure is dumb.

b) The special effects that such utterances are felt to achieve are lost if *but* (or some item signalling disagreement) is replaced by *and* in their second half; they are maintained, though, if the two clauses are merely juxtaposed:

(121) ?He may be a university professor, and he sure is dumb.

(122) He may be a university professor; he sure is dumb.

c) The modal bears contrastive stress, which seems to be crucial for the concessive overtones of the utterance. Interestingly, if *may* is omitted, the utterance can still achieve more or less the same range of effects in case the main verb is assigned contrastive stress:

(123) He IS a university professor, but he sure is dumb.

d) The proposition in the second clause is a contradiction of a contextual implication of the proposition embedded under *may* in the first clause. In (120d), the proposition that Jones is dumb obviously contradicts the implication of the embedded proposition 'Jones is a university professor', namely that Jones is intelligent.

The above observations are meant to suggest that the special effects that utterances like (120) are felt to cause (concession being maybe the most prominent among them) are not due to any special semantic features of the modal verb but rather to the specific nature of the material embedded under the modal, together with independent contextual considerations. It is useful to point out in this connection that such uses are not specific to English but are also exhibited by modal verbs expressing possibility in other languages. In Modern Greek, for instance, the verb *bori* denoting possibility has uses exactly parallel to those of *may*:[25]

(124)	Bori	na ine	kathigitis	panepistimiu,	ala
	May.3sing.pres.	3sing.pres.subj.	professor	of university,	but

	ine	anoitos.
	be-3sing.pres.indic.	dumb-masc.

Here is how a pragmatic account of these uses would go. As far as I can see, there is an important generalisation to be made about so-called speech-act uses: the material in the scope of the modal is (wholly or partly) metarepresentational. More specifically, speech-act uses involve metarepresentations of either linguistic material (utterances or parts thereof), or conceptual material (thoughts or parts thereof) which is typically attributed to a source other than the speaker at the time of utterance. I have already introduced metacommunicative and metacognitive aspects of the human metarepresentational device in section 2.4.1. In the following paragraphs, I will discuss their properties at greater length and deal with certain terminological points before applying them to examples of 'speech-act' modality.

[25] For some Modern Greek data, see Papafragou (1998b).

Consider what is involved in metarepresenting somebody else's utterances or thoughts (or, for that matter, our own utterances or thoughts): one representation is used to represent another representation (the original) which it resembles (thereby forming a second-order representation). In order for the metarepresentation to succeed, a certain degree of resemblance between the original representation and the metarepresented version of it has to be maintained. In metarepresenting propositional contents (either communicated or non-communicated thoughts), the content of one proposition can be used to represent the content of another proposition just in case they share a sufficient number of analytic or contextual implications (see Sperber and Wilson, 1986/1995: 227ff.). In other words, metacognitive and metacommunicative activity are constrained by quite general conditions on representation by resemblance. Following Sperber and Wilson, I have called the metarepresentational use of propositions based on resemblance in content 'interpretive use' of propositions (cf. the discussion in 2.4.1).[26]

Interpretive use is contrasted with descriptive use, where a certain proposition is used as a truth-conditional description of a state of affairs in the world. Consider the examples:

(125) Mary: So what did John say?

(126) Peter: It is a lovely day for a walk.

Peter in (126) might be taken to be describing a state of affairs in the external world; more plausibly, though, he is interpretively representing the content of John's utterance. The proposition expressed by (126) is thus used to represent what John said, and can be assumed to bear some resemblance to the original statement made by him. Since the utterance is used interpretively, what is important is not the truthfulness of the representation conveyed but its faithfulness to the original (see Wilson and Sperber, 1988b). Faithfulness is a matter of degree, so that metarepresentations can be more or less literal interpretations of the original propositional objects they are meant to represent.

The sort of interpretive use one may detect in (126) is not linguistically marked, and therefore rests wholly on the inferential side of communication. In other cases, interpretive use is lexically or grammatically indicated. Taking up the previous discussion in section 3.2, (127) is an example of overtly marked reported speech, where the metarepresentation is signalled by the adverbial *allegedly*:

---

[26] Apart from metacognitive and metacommunicative aspects, interpretive use also includes metalogical aspects; I am not concerned with these here (cf. sections 2.4.1, 2.4.3 and chapter 4).

(127) Allegedly, the Prime Minister will resign.

This is a case of *attributive* interpretive use, involving the attribution of a thought or utterance to some more or less specific source. This source may be indicated in very general terms, as in (127); it may be quite specifically indicated, as in (129a), or it may be linguistically unmarked, as in (129b):

(128) John: Flying will become more expensive.

(129) a. According to John, flying will become more expensive.
b. Flying will become more expensive, indeed!

Finally, on certain occasions, a metarepresentation is accompanied by an indication of the communicator's attitude towards the original representation. In (127), for instance, the speaker would often be seen as maintaining her distance from what the content of the metarepresented assumption conveys, whereas in (129b) the communicator may be seen as adopting an attitude of endorsement towards the metarepresented assumption. This subtype of attributive interpretive use of a proposition which includes expressions of attitude is called *echoic use* by Sperber and Wilson (1986/1995).

Apart from metarepresentations of propositional or conceptual content - utterances or thoughts - speakers may metarepresent aspects of linguistic form:

(130) On page 10, 'book' should be underlined.

Here, the speaker uses a linguistic stimulus to represent the form, rather than the content, of another linguistic stimulus. Linguistic metarepresentations which involve form instead of content are called *metalinguistic*. Together with the varieties of interpretive use (metacommunicative, metalogical and metacognitive), metalinguistic uses span the spectrum of metarepresentation in humans.[27]

On this account, cases of mention fall squarely within metalinguistic uses of language. When mention is accompanied by the expression of the speaker's attitude, metalinguistic uses are echoic in a way parallel to the interpretive echoic uses:

---

[27] This is not always the use reserved for the term 'metalinguistic' by the recent pragmatic literature (see Horn, 1985; Sweetser, 1990; Carston, 1998; cf. previous footnote).

(131) Johnny saw two 'mongeese'; somebody ought to tell him that the correct plural is 'mongooses'.

In the first clause of (131) the communicator metarepresents the specific linguistic form used by a previous speaker (a fact that can remain linguistically unmarked in spoken interaction) and (implicitly) adopts an attitude of rejection towards it; the second part of the utterance includes an explicit metalinguistic representation of form together with its explicit endorsement by the speaker, and thus functions as a sort of corrective statement with respect to the first clause.

We can now return to the examples of 'speech-act' modality. It is fairly easy to see that the material embedded under *may* in (109) is metarepresentationally used: as I showed, the utterance is felicitously used as a rejoinder to a previous utterance, which made the point that Jones is a university professor. Since the metarepresentational use here involves the content and not the form of the utterance, it is a case of attributive interpretive use. This is not overtly signalled by any linguistic device, but is quite directly retrievable by virtue of the immediate juxtaposition of the two utterances in interaction.[28] The same holds for the example in (110). What about the speaker's attitude toward the metarepresented content? I would suggest that (109) and (110) are not clear cases of echoic use, since there is no single communicated attitude towards the interpretively used material. On a first pass, it might seem that the speaker adopts an attitude of endorsement. However, she hastens to add that she disagrees with an easily accessible implication of the interpretively used content, which the initial utterance aimed at communicating: in (109) this is the implication that Jones, being a university professor, is intelligent; in (110) it is the implication that the interlocutors should have some beer.

The implication-denying aim of the second clause is bolstered by the existence of *but*. As Diane Blakemore has shown, this conjunct functions as a constraint on the implicatures of the utterance containing it: more concretely, it indicates that the proposition it introduces is meant to achieve relevance as a denial of expectation (Blakemore, 1989). In both (109) and (110), the second clause indeed contradicts and eliminates a highly accessible implication of the (interpretively used) content of the first clause. This is the reason for the awkwardness of (121): *and*, even after pragmatic enrichment, cannot achieve the same effects as *but*. However, a non-conjoined juxtaposition of the two clauses may convey effects quite similar to those achieved by the presence of *but*, as the example in (125) demonstrates. The reason is that the hearer can

[28] Of course, the interpretive use may be explicitly indicated, as in (i):
(i) He may be a university professor, as you say, but he sure is dumb.

inferentially work out the contrast between what is communicated by the material embedded under *may* in the first clause and the proposition expressed by the second clause - in other words, he can pragmatically supply the instruction which *but* encoded as to the direction in which the relevance of the second clause is to be sought.

On the present analysis, then, *may* in (109) and (110) has its normal epistemic interpretation. Its complement is an assumption which is derived by deliberate inferencing, and as such has come to belong to the speaker's 'belief-box' with a degree of strength attached to it. What is typical of so-called speech-act examples such as (109) and (110) is simply that the assumption in the complement of the modal has been picked up from the interlocutor's previous contribution to the exchange, evaluated and formed the output of the usual inferential computation of epistemic possibility.

The move of reducing speech-act modality in (109) and (110) to a subcase of ordinary epistemic modality has a number of advantages. For instance, it allows for the fact that very often the speaker chooses to modify her utterance with *may*, although she manifestly possesses enough evidence for the truth of the embedded proposition. In (109), the fact that Jones is a university professor is undisputed, so that (123) would actually be more appropriate given the speaker's current beliefs. By using epistemic *may* - i.e. by indicating that further evidence could conceivably bear on the truth of the embedded proposition, but that such evidence is unavailable to her at present - the speaker communicates less than what she manifestly knows. Therefore, she conveys that she does not want to commit herself to the whole array of cognitive effects the stronger proposition in (123) would produce. This strengthens the conclusion that the speaker does not subscribe to the contextual implication that Jones is intelligent. Moreover, the understatement gives rise to the implication that the speaker grants a point to the addressee now but reserves the right to dispute it later.

It may be objected here that, on my account, the speaker does not actually communicate in any strong sense that Jones *is* a university professor in (109). This goes against the intuitions of many speakers, and is explicitly argued against in Kay (1997). I have two lines of response to such an objection. To begin with, I do not think these intuitions are particularly decisive. For instance, they seem to rely heavily on the presence of *but*; compare (132a) to (132b):

(132) a. The minister may want to support research, but the funding is scarce.
b. The minister may want to support research; still, the funding is scarce.

More generally, it seems that the degree of speaker commitment to the complement of the modal depends on contextual assumptions, and cannot be derived in a purely structure-driven way. This is supported by the fact that there is a range of related uses of *may* which, although carrying concessive overtones, lack (at least on one of their readings) a speaker commitment to the complement of the modal:

(133) a. Harry may be a genius; I don't care. I never liked him.
b. The story of her life may be very sad - but I'm not interested in that. It's her work as an artist that I'm interested in.
c. He may be a university professor - but then again he may not; in any case, he's dumb.
d. This allegation may well be true, but it doesn't affect the rights of the defendant.

Roughly, the examples in (133) involve an *even if* type of concession rather than an *even though* type. In different terms, the speaker raises one possibility only to convey that it does not affect the main point she wants to communicate. In some of the utterances in (133) (e.g. (d)), it is difficult to distinguish between the 'speech-act' and a pure epistemic reading for the modal. This emerges more clearly in (134c), where the speaker is in no way committed to the sculpture being a masterpiece:

(134) a. X: I hate modern art. Look at this: what is it supposed to mean?
b. Y: Well, this sculpture is famous. It was the winner in the Royal Academy contest last summer.
c. X: It may be the masterpiece of the century, but I still don't understand it.[29]

How then are we to explain the various types of concession ranging from the strong variety in (109) to the weak version of (134)? Assuming that *may* has its usual epistemic interpretation throughout, it appears that the degree of speaker commitment to the complement of the modal (and hence, the degree of concession) will be pragmatically inferred. Generally, as I have suggested above, the presence of *but* encourages 'strong commitment' interpretations. In (109), by introducing the denial of an implicature of the complement of *may*, the presence of *but* supports the conclusion that the speaker actually subscribes to the truth of the complement ('Jones is a university professor') in the first

[29] One can also get conjunction-reduction on the two different interpretations of *may* (as Deirdre Wilson has pointed out to me):
(i) It may be famous, or indeed the masterpiece of the century, but...

place - and, therefore, grants a point to the addressee, who previously put forth a version of the complement. Another factor affecting the type of concession communicated is the resemblance between the original utterance or thought and its interpretive representation in the complement of *may*. In (134c), for instance, the complement of *may* is a less-than-literal interpretation of the original utterance of Y, and in fact an exaggeration;[30] rather than genuinely granting that the sculpture is a masterpiece, X communicates that, even if it is, he still denies that it is understood and recognised as such (which is a strong implication of the assumption that it is the masterpiece of the century). In fact, what (134c) communicates can be paraphrased as in (135):

(135) It is possible that it is the masterpiece of the century (although I doubt it), but I still don't understand it.

Yet a further advantage of considering speech-act modality as a subcategory of epistemic modality is that one can explain why *may* seems to have a privileged distribution in 'speech-act' examples over other modals - most notably, *can*. According to my semantic proposal in chapter 2, *can* encodes the information that the assumption in its complement (let us call it p) is compatible with the set of all factual assumptions that bear on its truth; on this analysis, *can* lacks epistemic interpretations. However, all concessive examples such as (109) and (110) appear to involve the speaker's admission that a certain assumption is not ruled out by her current knowledge - rather than by the state of affairs in the world. Furthermore, *may* is naturally suited to concessive examples, where the speaker temporarily grants a point while at the same time reserving the right to reject it later (as is especially the case in utterances such as (134)). This is so because changes in epistemic states are to be expected, so that what may now seem epistemically possible can in the future be disproved by further evidence (and vice versa). *Can* differs from epistemic *may*, in that no further assumptions entailing the negation of p (i.e. assumptions incompatible with p) can be uncovered in the future - hence the infelicity of *can* in the concessive examples of this section.

3.3.3 Further Examples

If the reanalysis of speech-act modality in terms of metarepresentations is essentially correct, then it should allow us to provide a unified account of a quite wide range of

[30] On exaggeration as a variety of representation by resemblance, see Sperber and Wilson (1989).

examples. On Sperber and Wilson's approach, what can be metarepresented may vary from actual utterances of previous discourse to implied propositions or unuttered thoughts, hopes, etc., and from fully propositional forms to individual concepts; furthermore, as I have already remarked with respect to example (134) above, the degree of faithfulness to the original proposition may range from total identity to weak resemblance. Let me explore some of these possibilities. Imagine that Peter and Mary are getting ready to go home after a party and the following dialogue takes place:

(136) a. Mary (to Peter): Give me the car keys. You are drunk.
b. Mary (to Peter): Give me the car keys. You are in no position to drive.
c. Mary (to Peter): Give me the car keys. You'll have an accident.
d. Mary gives Peter a worried look as he approaches the driver's door.

(137) Peter: I may be drunk, but I know what I am doing.

Peter's reply in (137) places within the scope of *may* conceptual material which has become available in the previous stage of his exchange with Mary. Offered as a response to (136a), Peter's utterance clearly interpretively represents a stretch of speech produced by Mary, and does so verbatim: in relevance-theoretic terms, it is a fully literal metarepresentation of the original utterance. If (137) is a reply to (136b), the material embedded under *may* bears a lower degree of faithfulness to the metarepresented utterance: in other words, it offers a less-than-literal interpretation of the proposition expressed by the second utterance of (136b), although it shares with the latter a number of logical and contextual implications (for instance, that Peter is in a bad condition, that he has no full control of himself, etc.). If taken together with (136c), (137) will be taken to interpretively represent a highly accessible contextual implication of Mary's second utterance ('You'll have an accident') - in particular, an implicated conclusion which Peter has derived in order to establish the relevance of Mary's comment. Finally, Peter may choose to interpretively represent an unuttered thought of Mary's, which he has reconstructed from the stimulus in (136d) - independently of whether the stimulus itself was ostensive or not; by interpretively representing Mary's thought Peter communicates that he has somehow retrieved it and wants to convey his attitude to it (as well as to one of its manifest implications).

Let me extend the analysis to some of Sweetser's earlier examples, repeated below for convenience:

(138) Editor to journalist:
OK, Peking can be Beijing: but you can't use 'Praha' for Prague.

(139) Mondale advisor giving directions to speech writer:
Reagan must be a nice guy (as far as the content of the speech is concerned), even if we criticise his policies.

(140) To smoker of long cigarette, from speaker who recognises that 'cigar' dialectally signifies 'long cigarette':
In New Orleans, you would be smoking a cigar right now.

Recall that metarepresentation may focus on aspects of linguistic form as well as aspects of content. I want to claim that, rather than warranting the postulation of a separate kind of modality, these examples simply contain metarepresentations of form. In (138), for instance, given certain pragmatic considerations, such as the authority relationship of the editor to the journalist, *can* is ascribed its usual permission-granting interpretation. *Beijing* is here used metalinguistically: the speaker is 'holding up' a term which (in all probability) was previously proposed by her interlocutor, and concedes its appropriateness for the identification of Peking. The proposition expressed by the initial utterance of (138) assumes roughly this form:

(138') Peking can be (appropriately) called 'Beijing'.

The existence of metalinguistic use is further supported by the overtly quotational use of '*Praha*' which follows. An identical analysis can be proposed for (140): the indefinite description *a cigar* is being used metalinguistically. The speaker refers back to a linguistic form which would have been considered suitable for picking out a specific object in a different dialect (that of New Orleans). In these examples one might further assume that the descriptions are used not only metalinguistically, but also echoically, since the speaker expresses an attitude towards their appropriateness to pick out specific referents. This line of analysis justifies Sweetser's feeling that there is something metalinguistic about these examples, and that they are possibly connected to the use/mention distinction. Moreover, it successfully accommodates cases like (138) and (140) to the metarepresentation of content I have examined above.

The case of (139) is slightly more complicated. On a first pass, it has to be set apart from (138) and (140) in that it involves no mention of linguistic form. What is distinctive about this example is the use of the proper name *Reagan* to refer not to the individual who used to be President of the United States but to a representation of that individual as it emerges from a text (here, a speech). Such uses of names and descriptions have attracted some attention in the linguistic literature; it is standardly assumed that 'real-world' referents are linked to their representations in images, pictures, photographs, etc. via 'pragmatic connectors', so that the natural way of talking about representations of a referent is by using the term for the referent itself (Nunberg, 1978, 1979; Fauconnier, 1985, 1997). In Fauconnier's theory of mental spaces, once a relation of the form 'x is a picture of y' is established through psychological perception, social convention, mode of production or some other means, then a local pragmatic connector is created to the effect that the term for y can be used to pick out the representation in x (Fauconnier, 1985: 12). In an earlier study, Jackendoff (1975) noted that this phenomenon is a pervasive characteristic of natural language, and its explanation should be sought not in 'imprecision' or 'loose metaphor' but rather in 'the means the language has to refer to images in pictures'.

Elsewhere I have raised some questions about the theoretical motivation and the descriptive adequacy of the concept of pragmatic connectors (Papafragou, 1995: 143ff.). For the moment, and without taking up the whole issue raised by pictorial representations and their linguistic expression, I would like to propose that (139) can be more profitably analysed using again the relevance-theoretic notion of representation by resemblance. Apart from whole propositional representations, isolated concepts can be held to enter into relations of representation by resemblance when they share logical or encyclopedic properties. Thus the concept REAGAN in (139) is used interpretively to refer to a representation of the concept's referent which is constructed out of assumptions included in a political speech. REAGAN, once used interpretively, no longer picks out an individual directly, but rather a description of an individual; even more accurately, an individual-as-object-of-description. Let's call the resulting concept REAGAN'. Of course, the new, ad hoc concept and the original concept for the individual resemble each other. After all, the derived concept has to maintain a reasonable degree of faithfulness to the parent concept if it is to be recognised as a representation of it in the first place. In addition to a common logical entry, the two concepts share a considerable part of their encyclopedic entry: for instance, a number (if not all) of the perceptual attributes for Reagan, as well as attributes concerning Reagan's personal history.[31] Where REAGAN and REAGAN' presumably come apart is in those attributes which have to do with

[31] I use 'attributes' for conceptual properties in the technical sense introduced by Barsalou (1992).

Reagan's political career and abilities. (139) explicitly conveys that REAGAN' must be a nice guy, even though Mondale's group criticises REAGAN's policies.[32]

This line of thought can be generalised to a broad range of cases where language is used to pick out pictorial or other representations. It seems that the possibility of using the term for a given referent to pick out a representation of that referent arises universally and spontaneously (Nunberg, 1978, 1979). This fact provides further support for a pragmatic account based on a general human cognitive ability, such as the ability to form representations by resemblance. A full-blown pragmatic explanation should also be able to accommodate the variations in the sorts of representation or in degrees of faithfulness to the original concept, the possible attribution of the representation to a specific source (i.e. the creator of a work of art), etc.

This may well be so, one might think, but what does it have to do with the sort of modality employed in (139)? Absolutely nothing - which is what I set out to show. *Must* in (139) may receive either of two interpretations. On a regular root interpretation of the modal, the utterance expresses a necessity in view of the circumstances; bolstered by further contextual considerations (e.g. the authority relationship of the Mondale advisor to the speech writer), the root interpretation may develop into a deontic one, whereby the speaker will be taken to impose an obligation on the hearer. In either case, there appears to be no room for a separate category of speech-act modality. (139) thus serves to strengthen the conclusion which emerges from examining all of the examples in this section: the class of speech-act modality is unified by the fact that modal operators may range over material which may be interpretively or metalinguistically used. This possibility, being a general pragmatic phenomenon, is not particularly linked to modality examples.

The proposed analysis can handle a number of formerly puzzling facts. Take first the observation that in utterances of the sort in (138)-(140) various modals seem to occur freely. An unrestricted distribution of modal verbs is exactly what one would expect if the occurrence of the interpreted/metalinguistic material is completely independent of the sort of modality involved, and is in fact a general pragmatic possibility. True, examples (138)-(140) are different from the other examples examined in this section. However, the reason is simply that they do not involve metarepresentation of the whole of the proposition embedded under the modal, but rather include metarepresentations of individual concepts (either as metalinguistic uses of form or as interpretations of content). This should not stand in the way of viewing all of the 'speech-act modality' examples as belonging to a

[32] Pronominal co-reference is not blocked by interpretive use (cf. *his policies* in (139), where *his* refers to real-world Reagan), although its acceptability may be influenced by a variety of pragmatic factors (see Fauconnier, 1985; Papafragou, 1995).

unitary group marked by the presence of metarepresentations. Natural connections between the two groups in (109)-(110) and (138)-(140) are not difficult to find. For instance, whenever the metarepresented concept is attributed to a previous speaker and endorsed by the communicator, the utterance in (138) acquires concessive implications parallel to those communicated by (109) and (110).

It is equally unsurprising that we cannot use a modal verb to modify an unexpressed description of the speech act performed by an utterance (e.g. an act of telling). What falls under the scope of the modal verb is the propositional content of the utterance embedded under the modal. Although this content may be descriptively or non-descriptively used, there is no room for a development of its truth-conditional content that would include a speech-act description which would then be picked up by the modal operator. As for linguistically expressed speech-act descriptions, again they do not warrant the postulation of a special sort of modality. In a case like (117) - repeated below as (141) - *must* has its normal root interpretation; in other words it expresses a necessity in view of the circumstances:

(141) I must tell you that your father wants you home.

Sweetser is right in pointing out the existence of restrictions on the co-occurrence of performatives and modals. But such restrictions can very often be predicted and explained on the basis of the cognitive effects the speaker might have intended to achieve by her utterance. It is true, for instance, that it is strange to assert - rather than demand - permission as in (115), repeated in (142):

(142) ?I may ask you where you are going.

Imagine now the following situation. Peter, Mary's husband, is getting ready to go out after having had a big quarrel with his wife:

(143) a. Mary: Where are you going?
b. Peter: You don't expect me to answer that, do you?
c. Mary: I may ask you where you are going. After all, I'm still your wife.

What Mary conveys by her reply is that she is allowed to ask Peter where he is going. The reason (142) is more acceptable within the context in (143) has to do with the fact that it creates a range of cognitive effects, i.e. it is relevant in that context. Assume, for example, that Peter manifestly wants his independence and therefore it is against his desires and preferences that Mary knows whatever he does at any given moment. It is easily inferred from the second assumption that it is also against his desires and preferences that Mary asks him where he is going. Given these manifest assumptions, Mary's response in (143c) communicates that she does not take into account Peter's desires and preferences in assuming that she is allowed to ask him about his whereabouts. There are two conclusions to be drawn from this example. Firstly, no special kind of modality is needed for the comprehension of (142). If the utterance is placed in the context of (143), *may* receives a normal root interpretation, i.e. it is understood to convey permission. Secondly, and more generally, there are no straightforward generalisations to be made with respect to the interaction between modals and performatives, since the acceptability of their combination within an utterance crucially depends on the way the utterance is intended to achieve relevance.

3.3.4 Metarepresentational Uses of Logical Operators

Let me add a couple of final considerations which extend (and indirectly support) the proposed reanalysis of speech-act modality. Firstly, the notion of interpretive/metalinguistic use should provide us with a means of capturing similarities between modal and other logical operators in natural language which operate on metarepresented conceptual material. Recent work within relevance theory has furnished two candidates: metalinguistic negation and some non-basic uses of indicative conditionals. Consider the following examples:

(144) a. He isn't neurotic OR paranoid; he's both.
b. I haven't DEPRIVED you of my lecture on negation; I've SPARED you it.
c. She's not my mother; she's my female progenitor.
d. The President of New Zealand ISn't foolish; there IS no President of New Zealand.

(145) a. If you eat TOMEIDOUZ, you must be from America.
b. If the wine bottle is half-empty, you are a pessimist.
c. If two and eleven makes thirty, you need more work on maths.

(146) a. If you're thirsty, there's beer in the fridge.

b. If I can speak frankly, he doesn't have a chance.
c. If I may say so, you're looking particularly lovely tonight.
d. Grandma is feeling lousy, if I may put it that way.

(144) presents some typical cases of what Horn (1985) has called 'metalinguistic negation' (see also example (115) above). This sort of negation is taken as communicating an objection to some property other than truth-conditional semantic content: the insufficient strength of the lexical item used in (144a), a non-truth-conditional aspect of the semantics of a word in (144b), the stereotypic assumptions or connotations that come with a particular word in (144c), or an existential presupposition carried by a sentence/utterance in (144d). Robyn Carston (to whom the examples and comments belong - see Carston, 1994: 322) has suggested that the correct generalisation about the uses in (144) is that the normal truth-functional negation operator takes scope over implicitly echoic material. Similarly Eun-Ju Noh (1998), after examining some allegedly non-truth-functional uses of conditionals, has concluded that the truth-table account of indicative conditionals can be maintained if the antecedents in (145) and the consequents in (146) are construed metarepresentationally (i.e. interpretively or echoically). Although there are various kinds of metarepresented material and various degrees of faithfulness to the original source, what unifies the examples in (144)-(146) is the presence of metarepresented form or content of actual or possible utterances/thoughts (or subparts thereof).

An account which allows for the existence of interpreted/metalinguistic material in the scope of logical operators has the advantage of keeping constant the operators' semantic contribution to the proposition expressed by the utterance, while offering a purely pragmatic explanation for the behaviour of the operators in metarepresentative environments. The opposite solution would be to adopt what Horn (1985) has termed 'pragmatic ambiguity', or 'built-in duality of use' for the logical operators. This move has been explicitly or implicitly adopted by a number of writers. As we saw, it should be considered to underlie Sweetser's discussion at least with respect to 'speech-act' aspects of modality (since she takes the root-epistemic distinction to be a semantic fact). However, as Carston (1994: 323) has remarked, what the 'pragmatic ambiguity' view entails is effectively a two-fold *semantic* ambiguity (see also Carston and Noh, 1995). On the one hand, there is a linguistically encoded ambiguity in the logical operators themselves, which is often manifest in the discrepancy between truth-functional and non-truth-functional behaviour. On the other hand, there is ambiguity in the nature of the material which falls under the scope of the operators, whether it is a proposition or an utterance. This two-fold ambiguity thesis is unsatisfactory on two counts. First, it is counterintuitive to postulate ambiguities for the logical operators. Second, the thesis is based on an odd redundancy, an unnecessary multiplication of ambiguities. The conclusion offered by Carston and Noh (1995: 7) is that 'this representational ambiguity is not a linguistic ambiguity nor even a pragmatic ambiguity, though it is a pervasive feature of language use. It is one manifestation of a perfectly general cognitive capacity of humans: the ability to metarepresent'.[33]

3.4 CONCLUDING REMARKS

In this chapter, I have taken issue with a range of arguments standardly used in support of (or in conjunction with) polysemy-based analyses of modality. Firstly, I showed that, claims to the contrary notwithstanding, the intended interpretation of a modal verb cannot be predicted solely on the basis of configurational properties but draws crucially on broader contextual considerations. Secondly, I argued that epistemic interpretations of modals contribute in a regular way to the truth-conditional content of the utterance; however, I presented some independent reasons for the fact that epistemic modals do not normally satisfy tests for truth conditionality, such as embedding under logical connectives. Finally, I examined Sweetser's category of speech-act modality, which was introduced as a natural extension of the polysemic organisation of the semantics of modals into the speech-act domain. My conclusion was that the phenomena subsumed under this novel category are not a natural class; consequently, cases of speech-act modality should be reanalysed as cases of either root or epistemic modality.

The discussion of the truth-conditional aspect of epistemic modality has made it possible to speculate on a fourth argument standardly used to support polysemy accounts for modals (but also for other lexical items): the argument from the historical development of epistemic modal meanings/interpretations in lexical items which originally expressed only root modality. The development of epistemic readings from root readings is assumed to be a diachronic semantic trend which indicates a rise in 'subjectification' in language (Traugott, 1995). I want to suggest that the link between epistemic modality and metarepresentation can lead to an interesting reinterpretation of the findings from diachronic developments. The Appendix to this chapter is an attempt towards such a reinterpretation.

---

[33] The same argument can be extended to all declaratives which can be construed either descriptively or interpretively: on the ambiguity view, even an example like (126) would come out as (semantically/pragmatically) ambiguous.

3A: APPENDIX

DIACHRONIC ASPECTS OF THE ROOT-EPISTEMIC DISTINCTION

According to what appears to be a robust cross-linguistic generalisation, epistemic modality historically developed from items which originally encoded other types of modal meaning (mostly volition, obligation and permission; see Bybee, 1988a, b; Bybee and Pagliuca, 1985; Bybee *et al.*, 1994; Bybee and Fleischman, 1995a, b; Heine *et al.*, 1991). This development has been linked, as I mentioned in 3.2, to a shift towards greater subjectivity in language, or the process of subjectification in grammaticalisation - in other words, 'the development of a grammatically identifiable expression of speaker belief or speaker attitude to what is said' (Traugott, 1995: 32; cf. Traugott, 1982, 1988, 1989).

To take just a couple of examples with *must* (from Traugott, 1989). Epistemic interpretations of the (root) modal first appear in the environment of a strongly epistemic adverb, such as *nedes* ('without doubt'):

(1) 1385 Usk, Testament of Love (Skeat) 109, 90:
He that dooth good & doth not goodly ... must nedes be badde.
'Whoever does good, but does not do it with good intentions ... must necessarily be bad'.

It is not until the seventeenth century that *must* occurs with an epistemic meaning in the absence of the adverb:

(2) 1632 Middleton, Spanish Gipsie I, i.16:
the fruit muste be delicious, the tree being so beautiful.

Several specific mechanisms have been proposed for the emergence of epistemics. I have already argued against the plausibility of one of them, metaphor (Sweetser, 1988; see section 2.1.2 and elsewhere). Another proposal includes generalisation, or semantic bleaching (Bybee and Pagliuca, 1985; Bybee *et al.*, 1994): on this view, epistemic modal meanings are more abstract/general than root ones. Although this solution is consonant with the general tendency in grammaticalisation for loss of meaning, it is not clear that epistemic readings of modals are indeed more abstract than root ones. Traditionally, bleaching is defined as loss of semantic features; a commonly cited example is that of the development of imperfective markers out of former progressive or habitual markers (Bybee and Pagliuca, 1985). Epistemic interpretations of modals, however, involve more than simply subtracting semantic properties (e.g. agency) from root modal readings and, in this sense, they are different from classic cases of bleaching such as that of the imperfective. In different terms, it is not explained why generalisation in epistemicity coincides with a development of the speaker's attitude towards the proposition embedded under the modal.

A third proposal invokes the conventionalisation of pragmatic inferences, or pragmatic strengthening (Traugott, 1989). For instance, permission readings in modals can be used to implicate expectation: if I allow you to do something (the claim goes), I expect that you will do it. Similarly, obligation may be used to implicate expectation: 'If one says *You must go* in the meaning 'You ought to go', one can implicate that one believes/concludes that it is true that you have to go' (ibid. p. 51). This pragmatic implication may then become grammaticalised and form part of modal semantics. Once the shift to epistemicity has been made, modal expressions can go on to acquire gradually stronger subjective overtones. However, I doubt that pragmatic strengthening can bring about epistemic interpretations, although it is definitely involved in other aspects of the historical development of the meanings of modals. The implication offered above for *You must go* does not really prepare the ground for the shift from root to epistemic modal readings; *must* still has its normal root meaning (as shown by the paraphrase *have to*). Moreover, obligation does not in fact imply inferred certainty. Finally, *You must go* does not admit an epistemic reading; more generally, epistemic and root readings of modals do not typically appear in the same environments (cf. 3.1), so the former cannot generally arise through pragmatic implication from the latter.

I want to argue that the metarepresentational analysis of epistemic modality, together with a modified version of the semantic bleaching hypothesis, can provide a satisfactory account of the historical change in the meaning of modals. Assuming a tripartite semantics for modal verbs, the emergence of epistemicity in a class of former root expressions can be explained as a change in the type of admissible modal restrictor. In the first epistemic uses of *must*, for instance, the specification of a root modal domain was dropped, so that the verb could range over metarepresentational material (a proposition used interpretively, rather than descriptively). The novel, epistemic uses of the verb were licensed by the fact that a number of inferences were preserved from the original semantic entry: the modal relation (necessity) was maintained but now it held between a metarepresentational assumption (the embedded proposition) and the set of the speaker's beliefs. After repeated uses, the semantic entry of *must* was reorganised, so that the former root restrictor was replaced by an unspecified restrictor which had to be pragmatically filled in. The output of this process was a more general/impoverished semantic entry for *must*. This prediction is consonant with the semantic bleaching hypothesis, but not as I have presented it so far: generalisation does not characterise the relation between the root and epistemic interpretations but between the initial, root reading and the final, indeterminate semantic entry for *must*.

On this account, the emergence of epistemicity in modal expressions is explained in two steps: firstly, the development of metarepresentational markers out of what were previously operators over descriptively used propositional material; secondly, the broadening/bleaching of the meaning encoded by former root modal items so that both root and epistemic interpretations can be derived on-line from an underlying semantic representation. This approach can be generalised to other modal verbs. In *should*, what has been described as a shift from 'weak obligation' to 'probability' can be explained as a shift in the normative restrictor encoded by the modal verb towards a metarepresentational domain. An interesting case is that of *may*, where the epistemic interpretation develops independently of the deontic (permission) interpretation (and indeed precedes it): both interpretations stem from the 'ability' meaning of the verb (Bybee, 1988a; Bybee *et al.*, 1994). This can be accommodated by an explanation which allows for the replacement of former factual restrictors with non-factual (metarepresentational) ones; however, it would not be easily accommodated by an account which derives epistemic possibility from permission through the conventionalisation of implications (as Traugott, 1989: 52 acknowledges), or through traditional bleaching.

A variety of linguistic items have followed a rather similar development. From contributing to the main explicature of the utterance (i.e. a conceptual representation which is responsible for the main bulk of cognitive effects caused by the utterance) these items have acquired uses which mark operations on the main explicature (thus contributing to non-basic or higher-order explicatures). Examples include the development of manner adverbs into evidentials (*clearly*, *obviously*) or illocutionary adverbs (*confidentially*, *frankly*), including 'stance' adverbs (*loosely*, *strictly*, *generally*). Not all of these items have identical properties synchronically, and all of them differ from modal verbs along a number of dimensions (see section 3.2); what unifies them is that they acquired metarepresentational uses at some stage in their development.

This type of diachronic development does not exhaust the content of 'subjectification'. It does not generalise, for instance, to the following cases (mentioned by Traugott, 1995): the development of the adversative/concessive out of the temporal connective *while*, the path from the imperative *let us* to the hortative *let's*, or the emergence of the 'preference' reading of *rather than* from a temporal reading ('sooner than'). But this is not necessarily an unwanted consequence, since it is not clear that 'subjectivity' applies uniformly to this larger class of linguistic items and constructions. Therefore, a metarepresentational account of the emergence of epistemicity (or 'subjectification') may be seen as a way of clarifying one aspect of what is essentially a term of convenience in historical linguistics.

This very brief discussion only touches on issues of diachronic modal semantics. I have mainly concentrated on the root-epistemic distinction with an eye to returning to the monosemy-polysemy debate on modality and reconsidering it in the light of historical evidence. Diachronic data have particular importance for polysemy analyses, since the presence of several secondary (historically older or newly emergent) lexical meanings is taken to point to a dynamic, cluster-like complex of encoded senses rather than to a unitary semantic address. In the case of the English modal verbs, the development of epistemics out of deontics is usually explained in semantic terms throughout. However, the directionality of historical change cannot be taken as evidence for the synchronic polysemy of modals for a number of reasons.[1] The first is that the directionality of historical change does not always coincide with the specific dependencies postulated by the synchronic analysis; I have in mind the case of *may*, where the epistemic reading historically precedes the deontic one. Secondly, and more crucially, the later appearance of epistemic interpretations of modal verbs does not exclude the possibility that, after an intermediate period of internal reorganisation, the semantic entry for modals underwent change towards a more general meaning (bleaching) which was capable of yielding a variety of contextual interpretations. Such a development presupposes that the semantic component operates in such a way as to maximise economy and avoid redundancy - an assumption which strikes me as reasonable. Instead of considering previously attested meanings as retained senses of a modal expression, this approach seeks to relegate to pragmatics the task of deriving the variety of modal interpretations in context. If this is

---

[1] I postpone the discussion of whether diachronic evidence is relevant for synchronic semantics until chapter 5.

right, then the overall diachronic precedence of root over epistemic meanings in modals does not straightforwardly translate into an argument for synchronically polysemic modal entries.

4

THE ACQUISITION OF MODALITY

4.0 INTRODUCTORY REMARKS

According to a widely held view in the psycholinguistic literature which has been around since the late seventies and early eighties, root modal meanings emerge earlier in language acquisition than epistemic ones. This observation has been accompanied by the implicit or explicit assumption that root and epistemic interpretations are separately specified in the lexicon; that is, developmental studies mostly took for granted an ambiguity/polysemy analysis of modality. In other cases, the acquisitional data have been used as an additional argument for a polysemy-based approach to modality (e.g. Sweetser, 1990: 50).

My aim in the following pages is to examine the evidence concerning the acquisitional priority of root over epistemic modal meanings, and outline some reasons for being sceptical about both its descriptive accuracy and its explanatory relevance to the semantics of the adult lexicon. With respect to the former, I suggest that it is not as robust a conclusion as an initial survey of the literature would lead one to believe: there seems to be at least some counterevidence showing that young children can and do understand epistemically coloured modal items (Hirst and Weil, 1982; Noveck *et al.*, 1996), or, similarly, mental terms (Moore *et al.*, 1989; Shatz *et al.*, 1983). As for the latter, the connection between developmental facts and polysemy-based analyses may be undermined if alternative explanations are found for whatever data genuinely show an earlier appearance of root interpretations. This can be achieved in various ways (and indeed it is plausible that complex factors will interact in the acquisition of a system as complicated as the English modal set). In the course of this chapter I will suggest that input data as well as performance limitations (mainly register inappropriateness) may influence the course of the child's acquisition of modality.

What I regard as the strongest explanation of the later development of epistemic interpretations, though, is my hypothesis that epistemic modality is metarepresentational.[1] If it is true that epistemic uses of modals mark the speaker's reflection on the content of her own mental states and thus hinges on abilities which are part and parcel of the child's developing theory of mind, then we have an independently motivated threshold for the emergence of epistemicity: the employment of meta-cognitive capacities is expected to reach an adequate level only well after the third birthday (and in fact close to four years of age). The metarepresentation hypothesis about epistemic modality, apart from its explanatory potential in dealing with developmental facts, meshes well with a unitary semantic account of the modals, as developed in the previous chapters.

The chapter is structured as follows. In 4.1, I reconsider the order of acquisition of the various meanings conveyed by modal expressions on the basis of evidence from either naturalistic longitudinal studies or experimental data. Section 4.2 explores alternative explanations for whatever data about the acquisition of modality I take to be genuine and illuminating, focusing in particular on the metarepresentation hypothesis. In section 4.3, I examine evidence from autism in the light of the previous analysis of epistemics. I conclude by sketching the implications for theories of semantic representation.[2]

4.1 Psycholinguistic Evidence: A Review

4.1.1 Naturalistic Longitudinal Studies

A number of studies focusing on syntactic aspects of modality (Brown, 1973; Kuczaj and Maratsos, 1983; Shatz *et al.*, 1986, among others, reviewed in Shatz and Wilcox, 1991) all point to the following general conclusion: the use of English modals begins gradually, between 1;10 and 2;6,[3] often with a single negative modal form (such as *can't*) appearing in limited syntactic environments (mainly declaratives). As noted by Shatz and Wilcox (1991: 331), modal vocabulary growth proceeds fairly rapidly during this early period, while the range of syntactic constructions in which the modals appear changes somewhat more slowly.

---

[1] This hypothesis was first discussed in connection with acquisition in Papafragou (1998c, f).

[2] I should note here that most of the psycholinguistic literature I am going to review uses the term 'deontic' loosely to refer to simple root uses as well; I will be consistent in my use of the term 'root' for both cases, unless the arguments bear directly and solely on deontic modality.

[3] The notation refers to years and months.

Other studies set out to provide evidence about the early semantic properties of the modals. Wells (1979) is a case in point. As part of the Bristol Language Development Study, he time-sampled 60 children along with their mothers every three months from 1;3 to 3;6. Wells' general finding was that epistemic modality is acquired later than root modality. He notes that by 2;6 more than 50% of the children used *can* to convey both ability and permission; by the same time, children used *will* to communicate intention. According to the same study, *could* emerges later than *can* and is used much less frequently for ability and permission; similarly, *may* is used for permission much less frequently than *can*. It appears also that by 3;0 *hafta*, *needta* and *wanna* are present with root meanings in child language (Gerhardt, 1991).

Wells (1985) reports on another sample of children followed from age 3;3 to 5;0. Between 2;9 and 3;0, children used *must, have (got) to* and *should* to communicate obligation or necessity, but, unlike *can* and *will*, these uses did not reach steady frequencies until later in development. In any case, by 3;3 all categories of root modality were in place. By contrast, children in his sample used *may* and *might* with an epistemic possibility meaning only by 3;3 years of age (here again, at least one occurrence from each child in half of the sample served as the acquisition metric). Use of modals to convey certainty was not achieved till much later, since by 5;0 only around 25% of the sample gave evidence of it. As for what Wells termed 'inferential' uses, e.g. the epistemic uses of *will* in examples such as *That will be the postman* (uttered on hearing footsteps), these seem to appear even later than expressions of certainty.

Shepherd (1982) notes a similar progression of meaning development in the preschool years - although she found quite frequent epistemic uses of *could* alongside the expected group of root uses. After studying a single child, she notes that *will* extends from intention/volition to prediction between 2;5 and 3;0; while *will* takes up the space of more distant future or of events in the immediate future which lie beyond the child's control, *gonna* emerges as the indicator of events in the immediate future which are controlled by the child (cf. Gee and Savasir, 1985, although their conclusions are also based on a limited number of subjects). In another study, Pea *et al.* (1982) found that, of 1,766 utterances containing a modal in the speech of a child between 1;11 and 3;4, only 7 express epistemic modality and 5 of these occur after 2;8. The discrepancy seems to extend to later ages: Kuczaj's (1977) results suggest that children between 2;6 and 3;6 produce in conversation more utterances with root modals than with epistemic modals when compared to children between the ages of 4;0 and 5;9 (cited in Hirst and Weil, 1982). Corroborating findings have been reported by Perkins (1983), who has also conducted some independent research on a large corpus of spontaneous conversation among 6-12-year-old children (recorded as part of the Polytechnic of Wales Language Development Project). Perkins points out that it is only later in development that children come to fully acquire the adult modal system, especially as far as the epistemic uses of modals are concerned.

Stephany (1979/1986) was one of the first studies to include acquisitional data from languages other than English - among others, Modern Greek, Finnish and Turkish. The results she cites are in accordance with the previous findings which placed the acquisition of epistemic modality in the second half of the third year, and in fact well after the third birthday (cf. Stephany, 1993). The developmental precedence of root over epistemic modality has also been observed in German (Stephany, 1993), French (Bassano, 1996), Polish (Smoszynska, 1993), Antiguan Creole (Shepherd, 1993) and Mandarin Chinese (Guo, 1994).

The univocality of previous research is challenged by the findings of Choi (1995). Choi studied the speech of three Korean children and traced the acquisition of five modal suffixes. The suffixes, which belong to an obligatory class of verbal inflections called sentence-ending (SE) suffixes, occur in informal interaction and are used to mark the status of the speaker's knowledge, i.e. evidential relations. More specifically, *-ta* indicates new/unassimilated information from the speaker's point of view, *-e* marks old/assimilated information, *-ci/cyana* indicates the certainty of a proposition (or shared information), *-tay* introduces non-shared knowledge/indirect evidence (e.g. hearsay or reported speech), and *-ta* (second type) marks information new to the hearer. According to Choi, this set of evidential suffixes is acquired between 1;8 and 3;0 years of age in roughly the order given above. The first two occur before the second birthday, *-ci/-cyana* appears productively with the onset of the second year, and the other two follow with three-month-intervals in between. Choi has also traced the acquisition of one epistemic modal auxiliary marking inference and of four root modals encoding obligation, desire, ability and permission: all of them coincide developmentally between 2;6 and 3;6 years, and in any case appear much later than the modal SE suffixes.[4]

Choi's research has been taken as *prima facie* evidence against the existence of a cognitive constraint on the early acquisition of epistemic modality (see, e.g., Shatz and Wilcox, 1991: 332). Before we go on to some explanations, however, there are some points to be made about the interpretation of her findings. Choi herself has suggested some possible factors which may facilitate the acquisition of Korean modal SE suffixes:

---

[4] In the same paper, Choi also reports on the use of the Korean modal auxiliary verbs in her sample. Of the five epistemic modal auxiliaries of the adult system, only *-na pwa* (indicating inference) appears productively (between 2;8 and 3;0) in the data. By contrast, around the third birthday, almost the whole of the root verbal system is in place.

their perceptual salience (given their sentence-final position), their obligatoriness (and thus the richer input they provide to the acquisitional device), and their semantic consistency (they do not communicate root meanings, neither do they incorporate tense or aspectual meanings). More importantly, she is concerned with the acquisition of an evidential system rather than a modal system based on the notions of possibility and necessity. Two facts are particularly important. Firstly, not all Korean modal SE suffixes were acquired by her sample. For example, children did not produce *-kwun* (:unassimilated inference based on newly perceived information). Secondly, not all the functions expressed in the adult grammar were present: *-ci* also expresses certainty of a proposition based on inference, but it did not appear with this meaning in children's speech. It is quite plausible that both evidentiality and epistemic modality turn on roughly the same *type* of cognitive abilities; in fact, I argue throughout that the latter - and possibly the former - are instantiations of the broader human metarepresentational capacity. However, epistemic modality makes stronger demands on the human metarepresentational device and it is therefore to be expected that genuine epistemic instances will emerge at more advanced stages of development.

It is important to stress at this point that, with the exception of Perkins, all of the above writers take a *semantic* root/epistemic distinction for granted; in other words, they use this distinction as a useful starting point for studying the acquisition of modal categories without worrying too much about finer distinctions of meaning. This often results in imprecision in the separation of individual root and epistemic meanings. As linguistic studies of modality have emphasised, it is often difficult to determine on a given occasion of utterance whether a modal verb should be assigned an epistemic or a root interpretation, especially in the absence of situational information (cf. the discussion of indeterminacy in Coates, 1983). The same argument applies with equal force inside each of the two broad types of modality. For instance, there seems to be little independent evidence for Wells' (1985) distinction between the 'inferential' and 'certainty' sub-types of epistemic meaning (as they appear, e.g., in epistemic uses of *must*).

Another point worth mentioning with respect to naturalistic data is that they can only yield results about language production. However, the link between production and comprehension is not as straightforward as it might seem. In fact, production in children as a rule lags behind comprehension (E. Clark, 1993); in other words, children avoid using parts of a linguistic system of which they already have a grasp until they feel quite confident in their linguistic knowledge. In an early study, Kuczaj and Maratsos (1975) demonstrated this very fact using elicited imitation with respect to the syntactic properties of the modals *can* and *will*. Inversely, very young children have been shown to occasionally use some members of a complicated grammatical system (e.g. an inflectional paradigm) correctly by simply memorising isolated items: production there does not guarantee successful acquisition of the whole system (as is further proven by the lack of overgeneralisations). It might be the case, therefore, that children do have a grasp of epistemic aspects of modality even before they start producing independent forms.

Let us accept for the moment (for what it's worth) the following fact: the data seem to converge on the point that the onset of epistemic modality follows that of root modality, and typically appears around or after the third year. Two types of explanation have been advanced in the literature reviewed above. The first has to do with input: it is reasonable to assume that most modal expressions produced by parents to children will have to do with permission, obligation, ability and other related notions, rather than with inference and the evaluation of the necessity and possibility of a conclusion. Although mentioned only in passing by the above authors, I believe this is a line well worth pursuing and I will return to it in section 4.2. The second type of explanation has to do with cognitive development, standardly viewed from a Piagetian perspective. For instance, Perkins (1983) attributes the early development of *can* and *will* in the preoperational stage to the child's egocentrism, which is also invoked to explain the absence of the more 'abstract' *must* and *may* (cf. also Choi, 1995). After 7;0, in the so-called concrete operational stage, the negotiation of social roles and tasks begins; this period coincides with the proliferation of expressions of root modality. Finally, from about 11;0 years (the formal operational stage) comes the abstract representation of alternative hypotheses and of their deductive implications, hence the productive use of epistemic modality.

Setting aside theoretical objections to the Piagetian analysis, which lie outside the scope of this chapter,[5] the basic problem with Perkins' proposal is that it gets the facts wrong. As most of the naturalistic studies have demonstrated, the first instances of epistemic modality appear around the third year, much earlier than a Piagetian account would allow. Anticipating the discussion in later sections, it can be argued on the basis of experimental studies that children already entertain and process logical possibilities by 7;0 years, if not earlier, thus lowering Piaget's estimated thresholds for logical reasoning in children (Piaget, 1928, 1987; cf. the similarly high estimates of Pieraut-LeBonniec, 1980). Moreover, evidence from the use of mental terms like *think, know, guess* to refer to one's own mental contents from around the age of 3;0 shows a type of cognitive ability which cannot be said to be tied to a simple, concrete mode of thinking.[6] In a nutshell, young

---

[5] For a starting point, see Piatelli-Palmarini (1980), Braine and Rumain (1983) and Wellman (1990). Cf. also Gopnik (1993: 7) for a snap argument against egocentrism.

[6] In this paragraph, I have related the Piagetian stages of development to specific ages of acquisition, as the linguistic (and the majority of the psychological) literature has tended to do. Smith (1993) argues that the

children seem to be capable of handling the concept of belief - and related epistemic concepts - much earlier (and in much different ways) than the Piagetian framework credited them with (for detailed discussion, see Perner and Astington, 1992). I conclude that, although the child's cognitive development is naturally the place to look for constraints on the acquisition of modality, existing linguistic/developmental analyses cannot adequately deal with the range of the child's emerging ability to talk and think about possibility and necessity.

4.1.2 Experimental Studies

The oldest and best-known experiments on the acquisition of modality (at least in the linguistic literature) are those conducted by Hirst and Weil (1982). 54 children between 3;0 and 6;6 were given two different modal propositions of varying strength (e.g. with a possibility vs. a necessity marker). In the epistemic cases, the propositions concerned the location of a peanut. In the root cases, they were commands by two teachers about the room a puppet was to go to. The child was to indicate in the first case where the peanut was, and in the second where the puppet would go. The strength ordering assumed was *is* > *must* > *should* > *may*.[7] The general result was that children appreciate the relative strength of epistemic modal propositions about a year earlier than root modal propositions (5;0 vs. 6;0 years approximately). Moreover, the greater the distance between the modals, the earlier the distinction is appreciated. Although these results seemingly provide arguments against previous research assigning to root meanings developmental priority over epistemic ones, the authors are careful not to draw any hasty conclusions. Note that, while the epistemic tasks were pretty straightforward, the root ones depended on the child's evaluation of the authority of the persons issuing the command, as well as of the puppet's compliance with the rules, so that performance rather than competence factors might have caused the root/epistemic discrepancy. What the experiments do demonstrate, though, is that the order of acquisition of the relative modal strength is the same in both epistemic and root tests. This result suggests that the two senses of modals have not

---

Piagetian framework is more appropriately seen as predicting broad differentiation over time, rather than being tied to specific age periods. Still, Smith admits that Piaget's own empirical accounts embody conclusions about the relations of children's age to developmental level, and notes: 'At best, this evident tension merits clarification; at worst, it marks inconsistency' (ibid. p. 104).

[7] Hirst and Weil ignored the possibility that an utterance containing *must*, on a logical/alethic construal of the modal, is actually stronger than an unmodalised declarative - probably thinking that such an interpretation would be unavailable to children of this age.

developed independently, and offers an indirect argument against polysemy/ambiguity-based analyses.

Byrnes and Duff (1989) largely replicated Hirst and Weil's study with children aged between 3;0 and 5;0 years. 5-year olds performed overall better than the younger age groups, while younger children performed significantly better on epistemic rather than root tasks. For reasons related to the performance factors mentioned above, these results are not conclusive: the authors used a highly desirable situation from the agent's point of view in the root tasks, and the children often responded using a rebellion strategy, that is, going against what they clearly understood to be an order. A more significant finding is that children performed better with words that were familiar to them (*has to, can't*) than more formal words (*might, might not*); although commonsensical, this fact has largely been overlooked in studies of the acquisition of modality.

Yet a third study inspired by Hirst and Weil's work is that of Noveck *et al.* (1996). Noveck and his associates initially replicated the original hidden-object task and checked the detection of relative (epistemic) strength contrasts in *is* > *has to* > *might* among 32 5-year-olds. The results largely confirmed Hirst and Weil's findings.[8] In a more controversial move, Noveck *et al.* assumed that epistemic modals like *might* and *have to* are not simply 'relative strength' indicators, i.e. they do not simply mark the speaker's commitment to the proposition which is embedded in the modal; these verbs also have a logical aspect, in that they mark possible or necessary inferences. They then went on to test the ability of children of various ages (5-, 7- and 9-year-olds) to appreciate the logical aspects of modals, and to compare their performance to that of adults. Subjects were presented with a hidden-object task, in which there were only two logically possible solutions. Subjects were in a position to arrive at these solutions on the basis of information they were given by the experimenters. In one case, e.g., they were given evidence which would lead them to form the assumptions 'There has to be a parrot in the box' and 'There might be a bear in the box'. The subjects then received two conflicting pieces of information from a puppet: namely, two modal utterances varying in strength, where the strong modal utterance was in fact misleading. In the above example, they would hear *There can't be a parrot in the box* and *There might be a bear in the box*. The experimenters predicted that, if children relied solely on the relative force of the modals,

---

[8] In fact, preference for *is* over *has to* was not significantly different from predictions based on chance. Various lines of explanation may be offered for this. Firstly, *has to* might be judged more natural (hence, more reliable) in a context where the hidden object has been out of view from the start; *is* would inappropriately suggest that the agent providing the hint possesses direct evidence for the location of the object. Secondly, there are many contexts in which *has to* is stronger than (entails) *is* - see previous footnote for a similar point about the *must/is* contrast.

they should agree with the hint carrying greater speaker certainty; if they were able to detect logical aspects of the modals, they should go along with the correct (albeit weaker) utterance. It turned out that children of all age groups performed overall correctly, with successful performance rates rising in older children.[9] 5-year-olds were better with logically necessary, rather than simply possible, conclusions, but again their scores were far better than those predictable by chance.

With their second experiment, Noveck and his colleagues have thrown some light on the use of logical inference and the evaluation of logical necessity and possibility by young children. In particular, the fact that 7-year-olds show a significant ability to detect logical possibility lowers previous estimates (going up to 9;0 years of age), although admittedly the experimental setting might have enhanced performance considerably. I will return to these results later on. For the moment, the issue is whether the second experiment has demonstrated anything about the comprehension of epistemic modality; I think it has not. There was, I submit, no difference in the comprehension of the modal utterances between the first and second experiment: both experiments involved epistemic interpretations. What did change was the range of evidence available to the subjects in either case. While in the initial series of experiments children had to rely exclusively on the puppet's hints to discover the location of a hidden object, in the second they also possessed independent information, which was apparently more reliable than any subsequent hints. Since the box containing the hidden object was closed from the start, it is reasonable to assume that the puppet supplying the modal hint might not have first-hand knowledge about the object's location, and in any case, it is not as reliable as the information objectively presented to the children before the trial began. The subjects were thus encouraged to draw on what they knew themselves, rather than give credence to the puppet's apparent guesses. Rather than revealing an early appreciation of the logical aspects of epistemic modals, the second experiment gives evidence of how epistemically modalised utterances can be used and evaluated as one information-source among others in young children's deductive reasoning.[10]

Let me finally turn to Coates (1988), who tested older children's comprehension of modals with respect to the epistemic/root distinction. She asked adult, 12-year-old and 8-year-old subjects to sort cards, each containing a modalised version of the sentence *I visit*

---

[9] In one variant of the task, equally strong modals were used, although only one of them was a valid hint: the prediction was that, if children rely on the relative force of modals, chance responses would occur. Even with 5-year-olds, though, the correct hint was chosen at a rate significantly greater than chance.

[10] See section 2.4.3 for reasons why Noveck *et al.* can't be right in conflating logical and epistemic modality. On the importance of metalogical reasoning for the full appreciation of epistemic modals, see 4.2.4.

my grandmother tomorrow, into piles on the basis of similarity of meaning. Cluster analysis revealed four distinct groups in the adult system:

(a) epistemic possibility (*may, might, perhaps, possible that, probably*);

(b) possibility/ability/permission (*can, could, nothing prevents, allowed, able, possible for*);

(c) intention/prediction/futurity (*will, shall, going to, intend*);

(d) obligation/necessity (*must, should, ought, have got to, obliged*).

12-year-olds also distinguished four categories, although there were some minor differences between the two subject groups as to the classification of specific items. 8-year-olds, on the other hand, did not include a category for epistemic modality and were in general less consistent in their options as a group. Coates concludes that the mastery of the modal system is achieved at a relatively later age. There are several potential problems with this experiment, however. First, a number of her examples are indeterminate between different interpretations. For instance, *I can visit my grandmother tomorrow* can indicate both ability and permission. Second, even the adult classification imperfectly matches the range of interpretations the modals may exhibit. *May*, for example, is capable of communicating permission as well as epistemic possibility, while *probably* can be said to convey prediction as much as epistemic possibility. Third, as Coates herself acknowledges, the linguistic construction she has used disallows certain interpretations, which are otherwise present in the language; logical necessity expressed by *must* is one example. Given that similarity is a fairly vague criterion itself, and that modals fall into various overlapping categories, it is not surprising that subjects seem to have classified modals depending on the verbs' preferred interpretations. It is, after all, pragmatic and not semantic intuitions that speakers typically tap when reflecting on the meaning of linguistic constructions (see the experimental data in Gibbs and Moise, 1997).

Overall, then, experimental evidence has been less illuminating than naturalistic studies as to the acquisition of root and epistemic aspects of modals. However, we still lack an explanation of these findings, and this is what I will start to develop in the next section.

4.2 A Developmental Proposal

4.2.1 Background

According to the hypothesis I have proposed in previous chapters, successful use of epistemic modals requires the speaker to perform deductive operations on abstract propositions (i.e. on the content of her beliefs *as such*) and to arrive at a warranted conclusion. As I have already mentioned in section 2.4, the metacognitive abilities which epistemic uses draw on have been examined in recent psychological work under the label 'theory of mind' (or 'folk psychology'). In what follows, I would like to construct an account of the acquisition of modal verbs by elaborating the link between epistemic modality and the child's developing ability to employ theory of mind.

Let me start with some background. Much current work has shown that children from an early age construe others as having mental states that underlie behaviour. For instance, it has been convincingly demonstrated that even 2-year-olds can distinguish between accidental and intentional actions (Tomasello and Kruger, 1992), and are sensitive to social cues such as eye gaze in detecting ostensive stimuli (Baldwin, 1991, 1993). Moreover, children at this age engage in pretend play which involves entertaining alternative conceptions of reality (Leslie, 1988). Around this time children start talking about perception and emotion (cf. the use of *see, look, taste, happy, like, want*; Bretherton and Beeghly, 1982).

Around 3;0, children start distinguishing between pretend and real identities of an object (Flavell *et al.*, 1987). They can also draw the distinction between real and mental entities, e.g. a real and an imagined cookie (Wellman and Estes, 1986). Three-year-olds are able to remember their own earlier desires (Gopnik and Slaughter, 1991). Moreover, they can predict a person's actions on the basis not only of what the person wants (this is something 2-year-olds can do) but also of what the person believes (provided these beliefs do not contradict the child's own beliefs; Wellman and Bartsch, 1988). Children of this age show some appreciation of the link between perception and knowledge: they understand, for instance, that one who looks into a box knows what is in the box, and one who doesn't look doesn't know (Pratt and Bryant, 1990).

For present purposes, I want to focus on some more sophisticated aspects of folk psychology (sometimes referred to as 'representational theory of mind'; Gopnik and Wellman, 1994; Leslie, forth.). These mostly involve the ability to appreciate that people stand in different and variable informational relations to the world - hence beliefs can vary, they may occasionally be false, and they are often modified or updated as new evidence becomes available. The most widely studied aspect of such advanced theory-of-mind capacities is false belief understanding. In one of the standard false-belief tests (Wimmer and Perner, 1983), a child is shown a character - Maxi, in the original version - who hides an object in some place and goes away. In his absence, the object is moved to a different location. Maxi comes back. The child is now asked to predict where Maxi will look for the hidden object. In order to answer correctly, the child has to realise that Maxi holds a false belief - that is, he/she has to contrast his/her own understanding of the situation with that of Maxi. This necessitates a more complex step than simply representing the state of affairs in the world: it requires the representation of someone else's (false) representation of reality. In a variation of the 'unseen displacement' task, the 'unexpected contents' task (Perner *et al.*, 1987), the child is shown a familiar container, such as a Smarties box, and is asked to predict the contents of the box. Contrary to expectation, it turns out that the box contains pencils. The child is next asked what a friend who has just come in will think is in the box. Again a correct reply relies on the child's ability to understand false belief in someone else.

It appears that children standardly fail false belief tasks before the age of 4;0 or 5;0. That is, most 3-year-olds respond to both the above tasks on the basis of their own (correct) beliefs. The conclusion that children around the age of 4;0 develop a better understanding of how the mind works is further evidenced by their performance in several related tests. For instance, in the 'unexpected contents' experiment, when 3-year-olds are asked what they initially thought was in the box, they reply 'pencils' not 'candy'; by contrast, children after 4;0 can accurately remember their own previous false beliefs (Gopnik and Slaughter, 1991). In a similar way, children after 4;0 begin to appreciate the distinction between appearance and reality (Flavell, 1986); for instance, they realise that something may look like a rock but is in reality a sponge. Around 4;0 or 5;0 children understand that people may acquire different information from the same perceptual experience depending on their previous knowledge (Taylor, 1988). They remember how they found out about a certain fact, i.e. the sources of their information (O'Neill and Gopnik, 1991). Moreover, they can link specific types of knowledge with the appropriate type of sensory channel (e.g. texture and touching, colour and seeing; O'Neill and Astington, 1990). Furthermore, it is only by 4;0 or 5;0 that children recognise that intentions are means to an end (Astington, 1991) and draw more reliable distinctions between intended and unintended actions which involve computing the agent's goals (Astington, 1990). After 5;0, children's understanding of the mind progresses in many domains: in knowledge about the mind/brain, in understanding embedded mental states such as beliefs about beliefs, and the social concepts that depend upon this, such as responsibility and commitment (Astington and Gopnik, 1991).

The interpretation of these empirical findings is surrounded by considerable theoretical disagreement. Some researchers take the wide-ranging changes in children's performance in theory-of-mind tasks around age 4;0 as evidence for conceptual change (Gopnik and Wellman, 1994; Wellman, 1990; Gopnik and Meltzoff, 1997; Perner, 1991); on this account, folk psychology is viewed as a truly theory-like object constructed through general-purpose learning mechanisms which resemble in crucial respect genuine (scientific) theorising. Other authors consider the 4-year-old's achievements as the result of improvement of independent performance factors (Leslie, 1991, 1994, 1995; Leslie and Roth, 1993; Fodor, 1992); this view assumes that theory of mind is a module underwritten by an innate mechanism. The arguments in this debate are extremely interesting but their evaluation lies outside the scope of the present discussion. It suffices that, regardless of how theories of folk psychology will turn out, they will have to take into account that several aspects of the acquisition of semantics/pragmatics presuppose the development of theory of mind. In this sense, the account of modality proposed in the next section does not hinge on differences among these proposals.[11]

4.2.2 The Theory of Mind Hypothesis

Against this developmental background, I suggest that the emergence of epistemic modal language is tightly linked to the advances in the child's performance with epistemological concepts which takes place around the third year. Recall that other properties which surface as a result of these advances are the ability to take into account the reliability of one's beliefs (i.e. the relative strength with which they are entertained), the ability to conceive of one's mental contents as representations distinct from reality, which may change over time or differ more or less dramatically from those of other people, and so on.

Significantly, one group of data which has been instrumental during the early stages of research on theory of mind consists of mental terms (e.g. *know, think, mean, remember, forget, guess, hope*, etc.). The acquisition of these terms, although distinct from, is obviously related to that of epistemic modal verbs, and might be useful in establishing a parallel between the two classes. More specifically, mental terms have been shown to arise in the spontaneous speech of children between 2;4 and 3;0 years of age (Shatz *et al.*, 1983); however, they are initially used rather as conversational devices (hedges), without

[11] It is an interesting issue whether modality could be used as a case study to bring out empirical differences between these approaches.

a full command of their representational meaning, in phrases like *you know, I think.* Genuine mental state uses begin after 2;8, often in situations where the child contrasts reality to a representation of reality (a belief, a dream, etc.). Other studies report that children are able to make distinctions of relative strength between verbs like *know/be sure* and *think/guess* only by their fourth year (Moore *et al.*, 1989, Moore and Davidge, 1989).

How can the empirical data on mental terms be brought to bear on the acquisition of epistemic modality? A first interesting correlation comes from the temporal pattern of acquisition. According to the data presented in section 4.1, occurrences of epistemic modality are rare before the third year, and mostly appear beyond 3;6 years. There is some indication that children use epistemic expressions such as *maybe* before modal auxiliaries (Stephany, 1979/1986: 395). This, however, can be viewed as an isolated rote-learned member of a system which the child cannot as yet fully comprehend - on a par with conversational uses of mental terms (cf. the memorisation of isolated parts of a complicated system in the first stages of acquisition, discussed in section 4.1.1).[12]

We have seen that after 4;0 years epistemic uses become more frequent and appear in a larger number of modal items. It is also at this period, and especially around 5;0, that children begin to understand strength distinctions among modal verbs, and between epistemic modal verbs and unmodalised declaratives (cf. the Hirst and Weil-type studies). These findings correlate pretty well with the mental verb data, and together fit the threshold of 4;0 to 5;0 for the solidification of folk psychological knowledge. Moreover, it makes good sense to assume that the ability to acknowledge the source of one's beliefs, to recognise one's past beliefs, and to realise that beliefs vary in strength and accuracy develop in parallel with one's ability to deductively process one's existing beliefs and to evaluate a proposition with respect to the cognitive support it receives from one's knowledge. A particularly strong piece of evidence for the present account is the fact that there is a correlation between the comprehension of the relative reliability of *know* vs. *think/guess* and the strength of epistemic *must* vs. *might* on the one hand, and performance on false belief, belief change and appearance-reality tasks on the other (Moore *et al.*, 1990).

Given at least the linguistic evidence, one would expect various aspects of metarepresentational performance to evolve across time and surface in different

[12] This deals with a potentially problematic point for my 'theory of mind' hypothesis: if theory of mind abilities are not fully employed before 4;0 years (and are still developing by the age of 6;0), how can the earliest uses of epistemic modal expressions be accounted for? My answer is this: 3-year-olds have some understanding of the mental world, which allows first (incompletely understood) epistemic uses. Even after 4;0, however, metacognitive performance develops over a long period of time, over which various aspects of epistemicity will emerge and stabilise. For instance, even within the space of epistemic modality, expressions of possibility precede those of necessity (cf. sections 4.2.3 and 4.2.4).

grammatical structures. If we go beyond the English modal verbs to other epistemic expressions, we would expect the development of the child's performance on theory of mind to place at least a lower bound on the emergence of hypothetical reference, conditionals, and other related constructions. It should be obvious that, far from attributing to young preschoolers the ability to perform *conscious* meta-cognitive reflection, my claim about their metarepresentational abilities concerns their growing ability to handle representations. In all of these cases, one must be careful to avoid circularity of argument, i.e. not to invoke the same type of data as both evidence for theory of mind and predictions following from folk psychological knowledge.

Still, one might ask, if the present status of the theory-of-mind hypothesis entails a threshold for the appearance of metarepresentational properties, how does it account for apparent counterexamples? I have already mentioned Choi's research on Korean SE suffixes, which has led some to question the existence of cognitive constraints on the acquisition of epistemic modality and to resort to weaker hypotheses (Shatz and Wilcox, 1991). Choi placed the acquisition of these evidential suffixes between 1;8 and 3;0 years, much earlier than a proper theory of mind can be said to emerge. Nevertheless, in what follows I want to speculate on possible ways of reconciling her findings with what we know so far about infant cognition.

First, as I have mentioned above, even 2-year-olds have some notions of internal psychological states; this is particularly evidenced by early 'conversational' interaction, facial imitation and joint-attention behaviour. Still, only simple causal relations are recognised between the mind (particularly desires and perceptions) and the world. The Korean suffixes acquired before the second birthday are *-ta* and *-e*, which mark new and old information respectively. A first guess then might be that, given the rich input they receive, young Korean children start marking off with *-ta* information that has recently been acquired through direct perception: this is borne out in Choi's data. *-e* is reserved for a variety of functions such as past tense, questions, etc. The appearance of *-ci/-cyana* in the beginning of the second year signals certainty of information, but again it is a certainty closely associated with direct and compelling perceptual evidence. Other uses of this morpheme are reserved for marking repetitions of a previous utterance or, in Choi's terms, 'shared knowledge' between interlocutors. However, such 'shared knowledge' uses do not exactly deserve their name, since they do not actually entail any representation of what the interlocutor *knows*; if true, this would indeed refute the present formulation of the theory-of-mind hypothesis. Repetition, though, need differ only minimally from marking something as old information. Consequently, *-ci/-cyana* need do little more than accompany a retrieval from memory. So far nothing in the Korean data necessitates a recognition on the part of the child of a mental state *as such*. This is not to say that the child does not reason about mental contents (e.g. propositions she has recently formed, or the sources thereof) but, crucially, she does not perceive these mental contents as *mental* contents.

What about the suffixes which occur in the third year? I think these are, in fact, more demanding. *-tay*, a reported speech marker, presupposes the ability to attribute to someone else thoughts, utterances or emotional states (corresponding uses of the suffix are documented by Choi). *-ta* (with a high pitch) is used by slightly older children for information which the child has from first-hand experience. Choi maintains that this suffix also conveys information new to the hearer: however, there is no evidence from her examples that it need be so (note that she says the same thing for indirect speech markers). What the utterances with *-ta* (Type 2) share is that they describe things a child might consider new, important or, in general, emotionally exciting, things which she wishes to 'brag about' (in Choi's words, who admits that there is an 'affective component' in what is conveyed by this suffix). This is consonant with an emerging theory of propositional attitudes, although it by no means implies that the child is in a position to detect partial knowledge in the hearer (an ability that normally emerges much later in development). Note, by the way, that if *-ci/-cyana* truly marked shared knowledge already from the second year, there should be no reason for the lag between that suffix and *-ta*, which would involve a comparable insight into the hearer's mental state.

Undoubtedly, the acquisition of modals and evidentials is not as straightforward as my discussion of Choi's findings suggests. One has to take into account formal properties of individual languages as well. For instance, Stephany (1979/1986) points out that modal categories such as the subjunctive in Modern Greek and the optative in Turkish appear earlier than the modal auxiliaries in English. The reason she gives is that morphological structural devices are acquired earlier than syntactic ones, since they are 'part of tightly knit lexical forms' (ibid. p. 198). Similar considerations apply in the case of Korean, where SE suffixes belong to well-formedness conditions for sentences, so there is pressure to start using them even if the child has not mastered the meanings of the morphemes. Moreover, I do not want to imply that the adult Korean modal suffixational system is definitely *not* metarepresentational (quite to the contrary). All I claim now is that young children may initially go about using this extremely complex system on the basis of a limited cognitive machinery, and no account has convincingly demonstrated that they need anything more to do so than what the theory of mind hypothesis already presupposes.

In fact, as I have hinted in chapter 3, I expect that most (adult) evidential systems draw on metarepresentational abilities to a large extent. There is some interesting evidence from Turkish which supports this claim. For all past tense expressions in Turkish there is an obligatory choice between two verb suffixes: *-dI* is used if the speaker was an eyewitness to the event; *-mIš* is used if the speaker has only indirectly experienced the event through hearsay or inference. Aksu-Koç and Slobin (1986) provide the following examples and glosses:

(1) Ahmet gel - di.
Ahmet come - dI
'Ahmet came dir/exp'.

(2) Ahmet gel - miš.
Ahmet come - mIš
'Ahmet came/must have come'.

The utterance in (1) is felicitous if the speaker has seen Ahmet arrive. The utterance in (2) is appropriate if, e.g., the speaker sees Ahmet's coat hanging in the front hall but hasn't seen Ahmet, while in the second, it is appropriate if the speaker has been told that Ahmet has arrived but hasn't seen him.

Aksu-Koç (1986) studied the acquisition of these suffixes by Turkish children. In one of her experiments, she showed children from 3;0 years up either illustrated stories in which a target event (e.g. the popping of a balloon) was explicitly shown or a sort of puppet show in which the event was hidden from the child but could be inferred from the perceived outcome of the story (e.g. the popped balloon). She reports that, when asked to relate the story, even children as young as 3;0-3;8 appeared to prefer *-dI* for directly perceived events and *-mIš* for inferred events. After 3;8 the reliability of the distinction in child language improved considerably so as not to differ significantly from that achieved by the oldest children in her sample (slightly over 6;0 years).

To test whether the use of evidentials was accompanied by genuine understanding, Aksu-Koç asked children to judge whether a doll which reported an event using either of the two suffixes had seen the event or was told about it. Performance in this task improved considerably after 4;0 years. Only 3 out of 24 children younger than 4;3 answered correctly, while all but one of the 36 older children passed the test. These results mesh very well with the theory-of-mind hypothesis for evidentiality: as I mentioned earlier on, the onset of advanced mentalising performance around the age of 4;0 is marked by an improvement in recognising the source of one's beliefs.

4.2.3 Auxiliary Hypotheses and a Sketch

In order to explain the specific order of appearance of root vs. epistemic modal interpretations in acquisition, we have to make use of more specific supplementary hypotheses about language acquisition, which will elaborate on the cognitive constraints placed by theory-of-mind development. A commonsensical, yet controversial, suggestion concerns the role of input in the priming of root over epistemic interpretations in the preschool years. Shatz *et al.* (1990) gathered samples of maternal speech to infants of around 3;0 years and found that fewer than 10% of the modals used had epistemic interpretations. Similarly, Wells (1979) remarks that *will* and *can* were used by all mothers in his sample, and they were also the most frequently occurring modals; we have seen that these two verbs were the first to be acquired by the children in the sample. Other studies focusing on semantic-pragmatic aspects of the acquisition of modality also detect a correlation between the input children receive and the type of modals they produce (Shatz *et al.*, 1989; Shatz and Wilcox, 1991).

Input considerations are crucial in the acquisition of word meanings, a process during which lexical entries are mapped onto conceptual addresses. Given a certain stage in cognitive development, input could partially determine the output of the child's production by triggering assignments of word meaning. In the case of modality, it is reasonable to assume that the increased frequency of certain root terms in the children's input is bound to facilitate their acquisition. I have in mind root interpretations which are within the cognitive grasp of young preschoolers and involve ability (*can*), desire (cf. volition in *will*), and later on permission and obligation.

A related argument comes from the frequently noted observation that epistemic modals, especially in English, are rather formal expressions, and would be unlikely to arise in children's spontaneous conversation (unlike other epistemic expressions like *have to*, or mental terms). An indication that register, alongside other pragmatic limitations, may influence the use and mastery of individual modal forms, even in the absence of competence-related obstacles, is the fact that *ought to* seems to lag behind the other modals (it is used to state obligation by only 3% of Wells' (1979) sample; Coates' (1988) subjects also had difficulty with the verb). Apart from the children's unfamiliarity with epistemic modals, it is also true that children normally find themselves more often in situations where the topic of conversation has to do with permission, ability, intention or obligation, so that even after the age of 3;0 they do not produce as many utterances with epistemic modals as with root ones.

There is obviously a lot more to be said on the interplay of these factors in the acquisition of modality, and a fuller picture of the predictions of the theory-of-mind approach remains to be given. Before moving on to an outline of this picture, however, I want to consider a possible objection to the analysis as it stands. It might be claimed that, on my proposal, the development of theory of mind seems to affect only one side of the root/epistemic distinction. This, however, is centainly wrong. Deontic modality, for instance, is concerned with whether something is necessary or possible on the basis of social or moral rules and regulations. Such rules are distinct from representations of reality, strictly speaking, in that they constitute descriptions of an ideal/exemplary rather than the actual world. Furthermore, other types of deontic interpretation rely on the desirability of a certain state of affairs, and thereby involve attitudes such as desires, preferences, intentions, goals and aspirations. Deontic concepts, therefore, seem to involve mentalising to a great extent. This has implications for language development. Suppose, following Premack and Premack (1994), that some basic knowledge about intention is not in place before 3;0 and develops even up to 6;0 years; suppose also that moral beliefs are not fully developed till even later. It follows that - the present literature notwithstanding - no genuine (adult-type) deontic modals can appear before these concepts are fully acquired.

It is definitely true that deontic concepts presuppose folk psychological concepts; moreover, they do so in more than one way. Notice, for example, that felicitous use of deontics often builds on complex social reasoning: in order to be able to use root modals to *impose* an obligation or to *grant* permission (and not merely note them), the individual must possess a range of assumptions concerning social relations, authority, etc. Therefore, I agree that there is an interesting story to be told about the acquisition of root interpretations, which will have to make reference to mental concepts. What is crucial, though, when comparing the acquisition of root and epistemic interpretations, is that the child may go about using the former without having a fully developed representational picture of mind, while in order for the latter class of interpretations to arise, this picture should by definition be consolidated in the child's psychology. The point is that the construction of alternative domains of assumptions (ideal, stereotypical, morally recommended, etc.) does not depend on metarepresentational capacities, since the individual does not have to focus on the representing act itself, or the content of a mental representation *as such*. Obviously, to the extent that some varieties of desire-based attitudes such as intention also rely on belief, the development of more specialised varieties of deontic concepts is tightly linked to metacognition. Still, we can maintain the presence of mature metarepresentational capacities as a general precondition for epistemic concepts and propose more detailed paths of development for more specific aspects of modal concepts (and their lexical expressions).

To illustrate, here is a sketch of how the development of modal expressions might go. The first uses of English modals appear before the third year. Consonantly with what a basic desire-intention psychology would predict, 'ability' *can* and '(quasi-) intention' *will* are the first modals to be employed to a significant degree by infants. Other desire-based predicates such as *wanna* and *needta* also appear at this period. With the emergence of a preliminary grasp of belief coincide other uses of root modals: *must* and *may* appear with increasing frequency at this stage to convey obligation and permission respectively. However, genuine deontic meanings cannot be communicated yet for the reasons given in the previous paragraph. When a child utters a statement containing (root) *must* at this period, she rather intends it as a description of a normative regularity (and similarly for *may*). This seems to be corroborated by the empirical data (cf. Perkins, 1983, who reports a predominance of non-performative, first-person-singular occurrences of modals at this age), and squares well with the sort of input the child receives from her caretakers.

Around the third year the use of epistemic modals and mental terms also begins, initially without full understanding of their meaning - these items are simply used as relative strength markers with no explicit representation of their inferential component. Further development of the child's folk psychology leads to the conception of the mind as a device which actively constructs and handles representations of reality. Developments of this type open the way to the comprehension of mismatches between one's mental constructs and the world (and, thus, to the identification of false belief, degrees and sources of belief, and so on). The child is by now capable of grasping the meaning of mental terms and of properly using epistemic modals like *may* and *might* (Wells, 1985). Between 4;0 and 6;0 the ability to calculate possibility on the basis of available data is solidified, together with an understanding of the differences in strength among modals depending on the type of inferential relation they encode (possibility or necessity). Still, a full grasp of the notion of necessity, which involves checking through all possible alternatives, escapes many young children. With regard to root interpretations, the active deployment of alternatives to the actual world, and especially ideal/morally recommended alternatives, together with a better understanding of the social consensus regulating issues of authority, duty, commitment and power, clear the way towards a proper use of deontic interpretations. Apart from register difficulties, the markedly later emergence of deontic *ought to* in acquisition can be attributed to conceptual difficulties associated with the domain of ideals or morality (cf. also Harris and Núñez, 1997: 220).

After the sixth year the child begins to acquire a fuller understanding of modal notions, particularly of necessity, which gradually proceeds to the full-blown development of the adult modal system (see next sub-section). On the semantic side, it is very plausible that, once theory of mind is fully deployed, the child will reorganise former lexical entries, and will trace the similarities between root and epistemic meanings based on common possibility and necessity underpinnings. By that time, the child will be able to construct a single lexical address for most modals and treat root or epistemic interpretations as pragmatic developments of an underlying unitary modal semantics.

4.2.4 Metacognition and Metalogic

Some further evidence for the account I have been advancing so far comes from the sort of modal notion which logicians have termed *alethic* (or *logical*) modality. As I mentioned in chapter 2, this type of modality concerns logical necessity and possibility defined independently of a thinking agent's mental contents, but in an absolute sense as relations between an (abstract) proposition and a set of propositions.

There seems to be some evidence for the late emergence of alethic concepts in children. Although some sensitivity to logical necessity has been attributed to infants as young as 3;0 (Fabricius *et al.*, 1987), it is generally acknowledged that children younger than 4;0 rarely take into account more than one possibility in hidden-object tasks (Sophian and Somerville, 1988).[13] It is also after 4;0 that an ability to declare a solution to a task undecidable between alternatives arises; it seems that, with training, children can recognise undecidability as early as 5;0 (Byrnes and Beilin, 1991). As other studies have shown, children at 5;0 or 6;0 have at least a receptive understanding of the necessary truth or falsity of propositions (Russell, 1982). More tellingly, though, the ability to reason about hypothetical possibilities, to generate possibilities that have not been specified in advance and to systematically collect and combine the information needed to move from a large set of possibilities to a single necessary conclusion is attributed only to much older children (7;0 to 9;0 years old - Sophian and Somerville, 1988; Byrnes and Overton, 1986; cf. Osherson and Markman, 1975). This ordering is mirrored in the acquisition of modal terms. Perkins (1983) reports that logical (what he calls 'objectified') interpretations of

[13] Memory limitations may also play a role in this (although for a different position, see Byrnes and Beilin, 1991).

expressions like *possibly, it is possible that, there is a possibility that* are acquired between 6;0 and 12;0 years.

Metalogical activities are typically conscious, or at least accessible to consciousness (Moshman, 1990). Although I have presented them separately here, I intend them to be construed as a domain closely related to metacognitive/metapsychological processes and developing in parallel with them. Although the situation is extremely complicated, one can complete the picture of the acquisition of modality offered in the previous section by venturing the following speculations. As part of the development of the child's second-order reflective capacities, and after the basic deductive abilities have been mastered, the child may tackle the complex metalogical task of reflecting on one's logical (deductive) processing steps. The latter capacity involves viewing propositional representations as abstract entities (thereby giving rise to alethic concepts). The development of metalogic, alongside that of metacognition, is accompanied by a deeper understanding of epistemic modal concepts, to the extent that it permits a full grasp of their inferential component. An initial manifestation of the comprehension of logical modality comes from recognising the solution to a problem as undecidable, when there is not conclusive evidence available - an ability which surfaces only around 6;0 years (Moshman, 1990). With the stabilisation of the ability to check through all possible alternatives and its generalisation across environments there emerges the concept of logical necessity, which is firmly established around 11;0 to 12;0 years. The child is then able to completely separate the logical and the empirical domain and to detect the validity of an argument based solely on its form, and not on the content of the premises. A number of researchers have recognised that, by this time, the child's use of the modal system closely parallels that of adults.[14]

4.3 A Note on Autism and Mindblindness

One thing to notice about the correlation between the emergence of epistemic interpretations and theory of mind is that it lends itself to experimental testing. An obvious place to look is autistic subjects' performance with epistemic modality. Autism is a severe childhood psychiatric condition which is characterised by a number of social and communicative impairments. Autistic children lack the usual cognitive flexibility,

[14] The discussion so far serves to clarify a possible confusion: one might argue that, since validity is standardly defined through alethic necessity, alethic modality is required for the successful use of root and/or epistemic modals. However, it is one thing to make a correct deduction, and quite another to understand its validity: only the first is involved in most early modal uses, and it requires logical, rather than metalogical, competence.

imagination and pretence, and their behaviour is marked by a restricted range of interests and activities.[15] Recently, a number of authors have suggested that (at least certain aspects of) autism can be explained as resulting from lack of the theory of mind, or mentalising capacity (Baron-Cohen *et al.*, 1985; Leslie, 1991; Frith, 1989; Baron-Cohen, 1995). On this hypothesis, autism is a form of mindblindness with a number of implications for cognitive development. Children who suffer from autism are unable to understand (and attribute) false beliefs; they cannot appreciate the mind/brain as an organ with mental functions; they are unable to realise that seeing leads to knowing; they experience difficulty with the mental-physical and appearance-reality distinctions. As I mentioned earlier in this chapter, onset of the acquisition of these aspects of the human theory of mind normally falls around the age of 3;0 or 4;0. Autistic children differ from both normal children and children suffering from various other disorders (e.g. Down's syndrome) in that their mentalising abilities are selectively impaired.

The mindblindness hypothesis about autism offers a solid testbed for the link which I have sought to establish between epistemicity and mindreading. If, as I have suggested, the development of the metarepresentational machinery responsible for mentalising is a prerequisite to the full-blown and correct production and comprehension of epistemic modal markers, then there should be a severe difficulty with such terms in the language of autism.

As a matter of fact, the available data are not always straightforward. de Roeck and Nuyts (1994) offer some evidence which appears to disconfirm this dissociation between autistic language and epistemicity. In their study, they investigate the use of three epistemic markers in Dutch by four high-functioning autistic adults: *waarschijnlijk* ('probable/probably'), *denken* ('think') and *kunnen* ('can/may'). On the basis of a corpus of spontaneous speech production, the authors conclude that all four subjects demonstrably use epistemic expressions in ways much similar to normal subjects.

Various explanations can be proposed for the success of the participants in the above study (the authors themselves explore some of them). It is possible to place these subjects together with the talented minority of autistic individuals (around 20-30%) who pass the (simple versions of) false belief tasks, and who have been shown to perform well in a variety of tasks involving metarepresentation. Such exceptional performance is usually explained in two ways:

(a) Autism might cause a delay in the operation of human mindreading capacities, rather than a permanent and absolute inability to metarepresent (Eisenmajer and Prior,

---

[15] For fuller descriptions of specific disorders in autism, see Wing (1976), Dawson (1989), and Baron-Cohen *et al.* (1992).

1991; Baron-Cohen, 1989, 1995). Depending on the different components which may be taken to form the human mentalising capacity, delay may affect one or several of them at different time-points.

(b) In high-functioning cases of autism, subjects may be capable of performing some sort of mentalising through the usage of alternative strategies. As Leslie and Roth (1993: 103) remark, 'such strategies are arrived at by exercising general reasoning abilities, and hence require a quite high level of such abilities, together with extensive practice and general knowledge, and hence do not appear before adolescence'. Although the exact nature of these strategies is at present not fully known, it has been suggested that they rely on an analogy to the pictorial format for the representation of thought. Pictures as a representational medium, unlike mental constructs, are rather well-understood by autistic subjects. Evidence for this is provided by good performance of autistics in variations of the false belief task which involve false (dated) photographs and false drawings (see Leekam and Perner, 1991; Leslie and Thaiss, 1992; and Charman and Baron-Cohen, 1992 respectively). High-functioning autistic people, then, may use their understanding of external representations such as pictures to achieve an understanding of mental representations such as thoughts and beliefs.

Both the age and the high verbal and non-verbal IQ of the subjects in de Roeck and Nuyts' study corroborate an explanation for their performance in terms of compensatory mechanisms for theory of mind abilities. In the same connection, it is worth noting that the youngest subject (and also the one with the lowest verbal and general intelligence) had markedly greater difficulty in using modal expressions than the rest of the group. de Roeck and Nuyts' study cannot, therefore, be taken to undermine the theory of mind hypothesis for autism or the metarepresentational analysis of epistemics.

A different strand of research shows that early autistic child language is quite impoverished in the use of mental terms. Tager-Flusberg (1993) reports the results of a longitudinal study, in which six children with autism aged between 3;0 and 6;0 were followed for between one and two years. Their spontaneous productions were compared to those of a Down's syndrome group of six children of the same productive language level and age. Her findings suggest that, while both groups of children talk about perception and mental states of desire and emotion, autistic children make significantly fewer references to cognitive mental states than children suffering from Down's syndrome. Elsewhere, Tager-Flusberg specifically comments on the use of modal verbs in the language of the two sample groups of the above study (see Tager-Flusberg, 1997). It appears that *can* and *will* make up 95% of all modals used by both groups of children, while epistemic uses are overall rare. What is more important for my present purposes, though, is that uses of these verbs which require some mentalistic understanding (e.g. the volitional/intentional uses of *will*) are very rare in the language of autism, whereas they occur more frequently in the speech of the children with Down's syndrome. In this study, impairments in theory of mind are reflected in the way language is used by children with autism. Moreover, it seems that this is done in a way consistent with the analysis of specific linguistic items (e.g. epistemics or volition markers) in mentalistic terms.

4.4 Concluding Remarks

In the previous pages, after having reconsidered the psycholinguistic evidence for the development of modality, I have advanced an explanation for the acquisitional priority of root over epistemic interpretations of English modals. More specifically, I have argued that the later emergence of epistemic uses can be predicted from independent assumptions about the child's developing performance in theory of mind. The elaboration of a link between epistemic modality and higher-order metarepresentational cognitive abilities yields two desirable results. On the one hand, it provides a motivated and theoretically exciting connection between early linguistic and conceptual capacities. On the other, it avoids the postulation of multiple senses for the modals in the adult lexicon. By maintaining a parsimonious, abstract semantics for modals, epistemic interpretations can be shown to arise whenever the speaker is taken to be deductively processing the contents of her beliefs *qua* representations in order to arrive at the proposition which forms the complement of the modal verb.

One might object that a theory-of-mind approach does not guarantee a unitary semantics for modals, and that some version of a polysemy account could be made to turn on the cognitive developments I have outlined. The fact remains that no such candidate analysis exists at present. Furthermore, a new-look polysemy account would have to abandon the central contention of (most of) its present counterparts that epistemic meanings somehow rely on root ones for their construction. In any case, no version of a polysemy account could deal satisfactorily with alethic modality. For instance, if the metaphor-based analysis were stretched to include alethic concepts, it would have to treat them as yet another semantic category, probably recognising some similarities between these and epistemic modal concepts. Such an analysis would have no means of motivating their late appearance in acquisition, since it is unclear how a novel metaphorical mapping could be used as part of the explanation.

5

Implications for the Semantics-Pragmatics Interface

5.0 Introductory Remarks

In the previous chapters, I have presented arguments against the view that English modal verbs are systematically polysemous and for the position that they possess a unitary semantics which contextually yields a variety of interpretations through general pragmatic principles. In this chapter, I want to abstract away from the specific points I have made so far and concentrate on more general issues pertaining to the psychology of word meaning. My aim will be to extend the arguments so far employed in discussing modality to shed some light on the problem of drawing the three-way distinction between monosemy, ambiguity and polysemy.

5.1 Options for Lexical Semantics: Monosemy, Ambiguity, Polysemy

5.1.1 Ambiguity vs. Monosemy/Polysemy

Faced with the problem of adjudicating between competing semantic analyses, one might in principle wish to maximise the amount of monosemy within the lexicon on purely methodological grounds. Semantic parsimony of the familiar occamistic type certainly militates in favour of abstractionist analyses. In view of this principle, polysemy and/or ambiguity should be restricted to cases where a monosemy account is clearly impossible. However, not all researchers working in the field would agree that the mere feasibility of a monosemy account guarantees its psychological reality; very often, there are further criteria which can be used to choose between equally feasible and descriptively equivalent semantic proposals.

As far as the distinction between ambiguity and polysemy is concerned, the most frequently mentioned such criterion is the intuitively felt connection between the various senses communicated by a polysemous lexical item - a connection which is absent in the case of ambiguity. In my analysis of modality, I have used this argument to shed doubt on the plausibility of an ambiguity-based approach to English modals. This criterion has some *prima facie* appeal, which is strengthened by certain experimental results. Subjects appear to produce judgements of sense relatedness with some consistency (Caramazza and Grober, 1976; Colombo and Flores d' Arcais, 1984; Durkin and Manning, 1989; cf. also next section). Nevertheless, there are various arguments which undermine its reliability. To begin with, introspection cannot give an accurate window onto either mental representation of word meaning or the way the entries in the mental lexicon are accessed during utterance comprehension. Moreover, similarity being such a vague notion, it is unclear what these intuitions precisely tap; it may well be a post-access stage of lexical comprehension, during which speakers assess meaning similarities on the basis of world knowledge rather than purely linguistic/semantic factors. Finally, inter-subjective agreement on sense relatedness judgements is not always univocal: Panman (1982) reports considerable variation in the responses of participants to a questionnaire-based rating of sense relatedness in polysemous examples such as *crime*, *break* and *neighbourhood* in (1)-(3) below:

(1) a. The crime had been discovered by a truck-driver.
b. He published several articles on crime.

(2) a. There was a break in the conversation.
b. There was a break in the water mains.

(3) a. There are several bookshops in the neighbourhood.
b. The neighbourhood objected to his plans.

So speakers' intuitions may be used only as preliminary evidence against an ambiguity analysis.

One way of sharpening intuitions of sense relatedness is by unpacking them into more concrete statements of semantic relations and overlap. One such statement is the following: If a word A consistently communicates two meanings, A_1 and A_2 in different contexts, then A may be considered truly ambiguous if there is no overlap between A_1 and A_2 (and, a fortiori, if A cannot communicate A_1 and A_2 simultaneously in any single occurrence). This criterion has been around at least since Weinreich's (1964) distinction between *contrastive ambiguity* (plain ambiguity, in my terminology) and *complementary ambiguity* (polysemy, in my terms) and seems to work for particular examples. Recall that in chapter 2, I argued that *can* in utterances such as (4) is not ambiguous between an 'ability' and a 'potentiality' reading because the former is, in fact, subsumed by the latter:

(4) a. John can give us a little lecture on cubism before we go to this Picasso exhibition; he knows so much about modern art.
b. John can give us a little lecture on cubism before we go to this Picasso exhibition; I'm sure he won't be too busy.

Similarly, in (5a) the two meanings ('aperture' vs. 'frame') of the polysemous entry *window* are collapsed (example from Pustejovsky, 1995: 48); in (5b) *see* can be interpreted as simultaneously 'perceive' and 'realise':

(5) a. John crawled through the broken window.
b. I saw that my father was still in the car.

No comparable examples can be given for genuinely ambiguous items such as *bug* ('device' vs. 'insect') or *cardinal* ('number' vs. 'priest'). The utterances in (6) are unacceptable on a 'collapsed' interpretation:

(6) a. *I find bugs very irritating and always try to remove them from my room.
b. *You have to pay attention to the cardinals: they are very important in the groups they belong to, be they priests or numbers.

More pressing arguments for keeping ambiguity separate from either polysemy or semantic indeterminacy come from standard syntactic tests. Here, I will mention only a couple by way of illustration (for more detailed discussion, see Zwicky and Sadock, 1975; Lyons, 1977: 405ff.; Kempson, 1977: 123ff., 1980, 1986; Kempson and Cormack, 1981; Cruse, 1986: 54ff., 1995; Atlas, 1989: 25ff.; Zhang, 1998). The first test relies on the possibility of co-ordinating the putatively distinct senses of a word in a single structure. Co-ordination in (7) is felicitous, thereby suggesting that *see* is non-ambiguous (i.e. it is either monosemous/semantically indeterminate or polysemous):

(7) I saw that the suitcase was empty and that Jones had lied to me.

By contrast, the zeugmatic feeling of (8) is a piece of evidence that the verb *expire* is ambiguous (example from Cruse, 1986: 13):

(8) *Arthur and his driving license expired last Thursday.

The second test involves the anaphoric expression *do so too*, the interpretation of which requires a same-sense antecedent. In (9), this requirement is met (example adapted from Kempson, 1977: 131):

(9) Brian killed a bird today and Jonathan did so too.

In (9), it is possible that Brian shot the bird, while Jonathan accidentally ran the bird over. This shows that *kill* is not ambiguous depending on whether the action denoted by the verb is intentional or not. By contrast, *run* in the following example fails the test: (10) cannot mean that Mary participated in the race, while Susan organised it. The impossibility of a crossed reading provides evidence for treating these two senses of *run* as belonging to distinct lexical entries with distinct semantic representations:[1]

(10) Mary ran the race and Susan did so too.

A final criterion for distinguishing ambiguity from polysemy/semantic indeterminacy involves the cross-linguistic persistence of the meanings systematically conveyed by a given lexical item. This 'argument from translation' draws on the following fact: In cases of ambiguity, there is an arbitrary connection of two or more meanings with the same surface form; as a result, one would not expect cross-linguistic similarities in ambiguity patterns. This prediction seems to be borne out. For instance, it is an idiosyncratic property of the English word *ear* that it is ambiguous between two senses: 'organ of hearing' vs. 'grain-holding part of a cereal plant' (example from Taylor, 1995: 103). The same grouping of meanings under a common form does not survive translation in different languages. By contrast, in both polysemy and indeterminacy there is some

---

[1] My use of 'lexical entry' in this chapter conforms to standard practice and should not be confused with the relevance-theoretic notion mentioned in chapter 1.

motivation behind the host of contextually manifested interpretations of a given term - viz. either a linguistically encoded relation (polysemy), or a pragmatically inferable connection (indeterminacy) between the corresponding concepts. Consequently, one might expect a certain degree of cross-linguistic agreement. Again, this conclusion is empirically vindicated. Examples include the root-epistemic alternation in modal terms, the use of the terms such as *mouth* and *eye* to refer to apertures other than the facial ones (as in *the mouth of the river*, *the eye of the needle*, etc.), and so on.

The above tests, when applied in conjunction, may lead to different predictions; indeed, the linguistic literature abounds with complaints about the variability of their outcomes. One reason for that is the involvement of non-linguistic parameters in constructing and interpreting isolated examples. Another is the genuine vagueness in some of the proposed criteria, such as sense similarity. Note, for instance, that for many speakers the two concepts expressed by *expire* in (8) are intuitively closely related; still, as the zeugmatic intepretation of the utterance shows, these two concepts are not part of the same lexical entry. I will have more to say about such clashes within lexical semantics in the following sections. For the moment, I conclude that, at least in principle, it is possible to draw the distinction between ambiguous and non-ambiguous cases, even though there may be disagreement over individual examples and interference of many factors in the application of diagnostic tests.

Assuming that the two phenomena are actually distinct, a number of researchers consider ambiguity as a less attractive option than polysemy for reasons of explanatory adequacy (see Gibbs, 1994; Taylor, 1995; for a different view, see Kempson, 1977, 1980). The justification offered for the tendency to maximise polysemy at the expense of ambiguity is that the former, unlike the latter, is internally transparent: i.e. the relations between the concepts encoded by a single lexical item can be read off from the semantic entry of the item. Apart from intuitions of relatedness, richer (:polysemic) lexical entries are claimed to account in a more natural way for novel uses of existing words (or sense extensions). For instance, one might propose that *climb* is ambiguous between the senses exemplified in (11); when predicated of animate beings, the verb denotes an act of ascending through laborious manipulation of the limbs, whereas otherwise it simply refers to a gradual ascent:[2]

(11) a. The girl climbed the tree.
b. The plane climbed to 30,000 feet.

[2] The polysemy-based analysis of *climb* is due to Fillmore (1982).

However, this is to miss an important generalisation: not only are the separate senses of *climb* related (through metaphor), they can also be extended in an equally motivated way to novel uses, such as (12):

(12) a. The soundtrack of the film climbed the charts.
b. I climbed into my suit.

The suggested conclusion is that *climb* and similar terms form clusters of inter-related meanings. Some of these meanings (e.g. the meaning of *climb* in (11a)) correspond to the prototype for the category, and are felt to be more central to the cluster as a whole. This model recognises that the processes which give rise to one-off, wholly creative novel uses of words are of the same type as conceptual processes (such as metaphor or metonymy) which relate the separate senses of a polysemous lexical item. Depending on factors of frequency, distribution, etc., former creative uses may go on to become conventionalised to varying degrees and thus become part of a polysemic cluster themselves.

The assumption that polysemy, unlike ambiguity, may account for the productivity of lexical meaning is shared by various authors in the field (Pustejovsky, 1995, talks in this connection of the 'generative lexicon'; cf. also Pustejovsky and Boguraev, 1996; Pustejovsky and Bouillon, 1996; Copestake and Briscoe, 1996; Ostler and Atkins, 1991; Nunberg and Zaenen, 1992). However, polysemy is not in any obvious sense a better candidate to explain lexical creativity than ambiguity. As the above tests for ambiguity have shown, it is often possible for two lexical items with an intuitive connection to come out as a case of ambiguity (e.g. *run* in (10)). In such cases, one may assume that the connection functions as a piece of etymological information which links two distinct lexical entries and does not participate in the synchronic, on-line comprehension of the corresponding word. Now the question arises: why can't this sort of etymological connection account for the fact that novel extensions are possible (and, indeed, pervasive) in word usage? I will explore this possibility at greater length in the next section, but for the moment I do not see what the advantages of polysemy over this construal of ambiguity are going to be.[3]

[3] By the same token, the generativity argument does not explain why polysemy rather than semantic indeterminacy is involved in the above examples.

5.1.2 Monosemy vs. Polysemy

I now turn to the division of labour between monosemy (semantic indeterminacy) and polysemy, which has been a central topic throughout this book. In contrast to ambiguity, both of these positions have the capacity to account for the intuitive connection between the various meanings communicated by a lexical item in different contexts. Faced with competing monosemy- and polysemy-based analyses, the question arises as to what possible criteria might be used to adjudicate between them. I will start with two, rather obvious, possibilities.

Firstly, monosemy is greatly favoured by the predictability of the range of lexical meanings through independent pragmatic principles from an underlying 'core' sense. Recall that, according to the account of modality proposed in the previous chapters, the deontic interpretations of modals are contextual developments of simple root interpretations rather than distinct semantic entities. For instance, the 'permission' interpretation of *can* in (13) is a pragmatic development (in the appropriate contextual conditions) of the 'potentiality' meaning which the verb grammatically encodes:

(13) You can leave the door open.

This analysis of deontic interpretations bears similarities to the relevance-theoretic analysis of mood (Wilson and Sperber, 1988a, b). On the original Wilson and Sperber analysis, a variety of mood indicators may take on 'performative' overtones which are independently motivated by the principle of relevance and broad conditions on contexts. Within this framework, expressions encoding potentiality, such as the subjunctive in Modern Greek (Rouchota, 1994a), can be used to convey permission in a quite general way, depending on assumptions about status and authority among the interlocutors (see also Clark, 1991). Another example where pragmatic/conceptual factors are responsible for systematic meaning multiplicity associated with specific linguistic constructions is the metalinguistic use of modals, negation and conditionals which I have surveyed in an earlier section. The presence of an independently established pragmatic phenomenon, viz. the ability to set up and comprehend a specific type of metarepresentation, relieves the semantic component of the need to account for these meaning variations in a principled way.

Secondly, monosemy is favoured by the absence of separate grammaticalisation/ lexicalisation of the candidate senses in different languages. The more likely it is for a certain process to be pragmatic/natural, the less likely it is for this process to be formally marked. This is the inverse side of the argument already put forth for distinguishing polysemy from ambiguity: true ambiguity is placed very low on the predictability scale and is expected to block translation equivalences across languages. By contrast, clear cases of univocal semantic entries which give rise to systematic, pragmatically determined interpretations maintain cross-linguistic stability. Examples are offered by the metalinguistic uses of logical operators; especially for the better-documented case of negation, it is known that metalinguistic uses are widespread but remain cross-linguistically grammatically unmarked (Horn, 1985). Similarly, the pragmatically enriched versions of the *and*-conjunction in (14) do not seem to receive distinct encodings, although they are attested in a variety of languages (see Carston, 1998 for discussion):

(14) a. She ran to the edge of the cliff and jumped. (and then...)
b. I insulted him and he resigned. (and as a result...)

Although the monosemy end of things is rather uncontroversial, the argument from grammaticalisation is less secure on the polysemy side. It is standardly assumed that, as far as grammaticalisation is concerned, polysemy is different from both ambiguity and monosemy. Although a degree of uniformity persists across languages, there are differences among separate grammatical systems concerning which dimensions of a broader conceptual space are going to be encoded. Indeed, this has been one of the main arguments in support of a polysemy-based analysis of the English modal system (Sweetser, 1986). On this view, modal expressions in different languages carve up the broad conceptual space of modality in different ways; as a result, there is no straightforward equivalence between the modal systems of various languages. Moreover, given that the epistemic/root distinction is a major structural divide within this conceptual space, it is not the case that all modal expressions systematically alternate between epistemic and root interpretations. It is therefore concluded that the root/epistemic distinction in separate modal systems (or in specific lexical items within a modal system) is the output of linguistic encoding and not pragmatic processing - hence of a polysemy-, not a monosemy-based semantics.

I do not find this argument for polysemy particularly compelling. As far as the example of modality is concerned, my analysis also views it as a broad conceptual space within which commonalities as well as individual variations are expected among grammatical systems. However, it has been my contention that the root/epistemic distinction is not necessarily grammaticalised within the conceptual space of modality;

rather, some aspects of the distinction are understood through very general pragmatic processes which draw on our capacity to metarepresent. This account does not overlook the fact that particular languages place specific grammatical constraints on the modal system so that individual modal systems (or modal expressions) may differ along a number of parameters. Indeed, this is one of the reasons why a given modal item may be semantically univocal, although different from superficially similar modal expressions in other languages (e.g. compare English *must* to German *muß*).

One may object that a purely pragmatic account of systematic meaning multiplicities runs the risk of overgeneration. To revert to modality, not all English modal verbs which have root interpretations also admit epistemic readings - *can* being the obvious counterexample. However, such non-occurrences can be predicted even within a pragmatic account in a number of ways. One solution is through blocking by semantic constraints: I have argued above that the type of admissible restrictor for *can* precludes the possibility of epistemic interpretations. Other cases may be explained through pre-emption by existing members of the grammatical/lexical system.[4] For instance, one may argue that the animal/meat alternation in pairs such as (15) is a pragmatic (rather than a semantic) phenomenon, i.e. it is a case of inferential derivation rather than polysemy. One may then predict restrictions on the alternation in examples such as (16a) on the basis of already available lexicalised options such as (16b):

(15) a. I saw a rabbit in the garden.
b. We had rabbit with onion sauce for dinner.

(16) a. *Sheep with potatoes is a traditional Greek roast.
b. Lamb with potatoes is a traditional Greek roast.

If this is right, it follows that the argument from translation is less than absolute: it is perfectly possible to have univocal lexical items in a certain language which lack exact translational equivalents in other languages. Monosemy does not guarantee universality, because it is shaped by what is grammatically possible in different linguistic systems. Seen in this light, modality does not present a puzzle: the lexical meanings of modal expressions are expected to cross-linguistically overlap only partly (if at all) in terms of the modal relations and modal restrictors they encode. But if we cannot expect monosemy

[4] In fact, pre-emption may also be invoked in the case of *can*: *may* can already express epistemic possibility in the English modal verb system.

to coincide with universality/translatability, we cannot use the latter as a criterion to distinguish monosemy/semantic indeterminacy from polysemy.

Are there any other ways of distinguishing between these two options? Given competing analyses, a large number of authors consider polysemy-based semantics as preferable to monosemy (and ambiguity) alternatives for several reasons. Some involve diachronic and developmental arguments; I will leave these for a later stage in the discussion. In the remainder of this section, I will concentrate on some synchronically relevant facts which have been presented as obstacles to monosemy analyses, and hence as reasons to opt for polysemy-based semantics (Taylor, 1995; Sweetser, 1986; Gibbs, 1994; Rice, 1992; for a different position, see Ruhl, 1989; Bierwisch and Schreuder, 1992; Fodor, 1998; Fodor and Lepore, 1998).

Firstly, it is often noted that abstract meanings for items which exhibit a rich array of superficial meanings are generally hard to state. For instance, Taylor (1995) in his discussion of *climb*, notes that the central sense of the verb (exemplified in (17a) above) can be paraphrased as 'locomotion from a lower to a higher level by means of a fairly laborious manipulation of the limbs' (ibid. p. 106). However, this paraphrase cannot be extended to capture the rest of the meanings communicated by the verb. Even if one drops all the specific attributes of the core meaning and assumes that *climb* semantically encodes 'ascend', the paraphrase 'scarcely does justice to the complexity and subtlety of the individual senses'; moreover, it fails to 'provide any adequate basis for distinguishing *climb* from other verbs which also profile the upward movement of an entity, such as *ascend, rise, go up*, etc. ' (ibid. p. 107).

Here, as elsewhere, a monosemy position is seen as involving some sort of decompositionalism. There is a requirement on the core meaning of a lexical item not simply to be effable in natural language but also to possess a sufficiently high degree of effability (so as to surface in paraphrase). This conception of the core lexical meaning is reminiscent of the classical model of categories, which attached to the great majority of natural-language words a set of individually necessary and jointly sufficient conditions for category membership (i.e. a definition). As I have mentioned earlier in this chapter, the decompositionalist program broke down because of insoluble empirical and theoretical problems, not the least of which was the impossibility of formulating satisfactory definitions for most natural-language items. If the semantic entry for indeterminate lexical concepts is assumed to be a different concept, which somehow constitutes the core or basis for the contextual development of a range of lexical interpretations, then for many concepts this project will reiterate the difficulties of semantic decomposition.

It is not the case, however, that semantic indeterminacy requires semantic decomposition. In certain instances, it may be possible to be precise about the concept encoded by a lexical item which exhibits a variety of contextual interpretations: in the case of the *and*-conjunction or the negation, there is a logical vocabulary which furnishes candidates for underlying senses. In other instances, univocal semantic entries are not paraphrasable without loss or distortion of meaning. This is not an evasion on the part of the theory: even if it is not possible to specify all the conceptual attributes for a certain lexical concept, it is certainly feasible to isolate the dimensions along which the concept is indeterminate. For instance, it is easy for speakers of English to identify the aspects of the meaning of *fast* which are unspecified by its semantic content but will be filled in through contextual modulation of its meaning:

(17) a. John's car is fast.
b. She is a fast typist.
c. Avoid fast movements.

A second argument against monosemy and in support of polysemy comes from speakers' ability to assign degrees of relatedness to surface interpretations of lexical items in non-symmetrical ways. Consider the example of *line*, which can contextually communicate a range of meanings. Caramazza and Grober (1976) classified these meanings into clusters according to the intuitions of subjects in a number of experiments they conducted. Below I give the central (:most frequent and typical) sense of *line* in (18) and two of the clusters in (19) and (20):

(18) a. We were told to line up.
b. The shortest distance between two points is a straight line.

(19) a. Ford is coming out with a new line of hard tops.
b. He had come from a line of wealthy noblemen.

(20) a. When the curtain rose for the second act, Bob could not recall his opening line.
b. She said it was a line from Keats.

Informally, *line* in (19) denotes a sequence/ordering of constructs, while in (20) it denotes a continuous sequence of words. On a polysemy-based analysis, these might be considered independent (although connected) senses of the word; a monosemy-based analysis, by contrast, would take *line* to be univocal but drawing on distinct contextual assumptions in each example. A potential problem for a univocal semantic analysis is to explain the relative distance between the various contextually-determined meanings in a way that respects the above clusters.[5]

Again I do not think this argument presents a serious challenge to the monosemy view. One way of accounting for the cluster effects is to consider them a product of pragmatic processing, brought about by the relative accessibility of intermediate contextual assumptions which are necessary for the derivation of contextual readings. For instance, both the utterances in (20) require the activation of encyclopedic assumptions about theatre plays or poetry, the form of their written versions, etc.; the geometrical concept of *line* given in (18) can thus be used loosely to refer to a unit of written discourse.[6] The situation is different in (19). In (19a), for example, the linguistic content of the utterances activates assumptions about the way cars are produced, the way identical models come out of the factory in a (geometrical) line, etc.; this context licenses a rather different interpretation of *line* from the one in (20). So although pragmatic derivation rather than semantic decoding is responsible for the derivation of the different readings of *line* in (19) and (20), we can still predict differences in the way these readings are interrelated.[7]

In the same connection, it is sometimes observed that creative sense extensions normally apply to a more specific sense of a word (i.e. to a part of a polysemy cluster) rather than to a broad, underlying meaning. To take the example of *line* again: it may be claimed that the metaphorical use of (21) is derived from the more specific sense in (18) - rather than the abstract, geometrical sense of *line* - since in both cases the noun denotes a series/succession of constructs. This would conform to the cluster organisation of surface meanings in the experimental setting, in which examples such as (19) and (21) were grouped closer to each other than to either (18) or (20):

(21) We came to different conclusions using the same line of reasoning.

This type of argument does not unequivocally point towards polysemy. As before, the relatedness of different interpretations can be explained as a result of the parallel

---

[5] This is not a problem according to Caramazza and Grober (1976); but see Taylor (1995: 285).

[6] On the loose use of concepts, see Sperber and Wilson (1985/6, 1986/1995), and chapter 1.

[7] Closeness to the prototype for the category is also responsible for the clustering of interpretations of *line*. However, this is not incompatible with monosemy accounts: one can assume a unitary semantics for *line* and accommodate typicality effects as products of the organisation of the encyclopedic entry of the concept. One prediction that does *not* follow from the monosemy account (although it is entailed by some approaches to polysemy) is that the basic or core meaning of a term will also be invariably perceived as the prototypical one. This is a good thing, as section 5.1.3 will show.

activation of different encyclopedic assumptions in the construction of a pragmatic interpretation. The intuitions of relatedness will come out as a post-access statement of the similarity between different contextual interpretations of a single lexical entry (where similarity is defined in terms of shared logical and contextual implications).

A final point used to support a polysemy over a monosemy analysis is the fact that one of the meanings systematically communicated by a lexical item may be concrete, and hence more basic. This way of formulating the argument begs the question, since it presupposes a questionable distinction between concrete and abstract concepts. More specifically, it is not clear what the abstract/concrete distinction buys in explanatory terms: in the example of *line*, is the 'geometrical figure' interpretation (which speakers judged to be basic/prototypical) more concrete than the interpretations required for examples (18) or (20)? If not, is this a counterexample to the theory? Finally, how does *line* differ from predicates such as *exist*, which may felicitously apply to different items without a change in meaning? In (22), slightly different concepts for *exist* are retrieved for each utterance; this, however, can be attributed to the (appropriate) difference between God and grizzly bears, not to two separate senses of the verb *exist*:

(22) a. God exists.
b. Grizzly bears exist.

The point is that, depending on the sort of things which form a (geometrically defined) line, the concept retrieved for the comprehension of *line* is slightly different: lines of trees, lines of argument and lines of car all differ along this dimension. We do need a (pragmatic) theory about how this contextual modulation of the concept encoded by *line* takes place (and I will make some suggestions in section 5.2 below); but I don't think we need a semantic theory (involving polysemy) for all the different concepts of line we can construe and communicate.

Let me summarise the argument so far. I have outlined certain criteria which are meant to separate polysemy from either ambiguity or semantic indeterminacy. The results suggest that, although in principle the three phenomena are distinct, there is some difficulty in applying specific tests and obtaining univocal results. Furthermore, a range of considerations which have been assumed to favour polysemy-based analyses over the other two options have been shown to be inconclusive. Consequently, although there are some genuine cases of systematic meaning multiplicity which are not captured by either ambiguity or monosemy accounts, this class may be much smaller than the range of phenomena which have been subsumed under polysemy in linguistic analyses. In other words, the scope of polysemy in linguistic semantics is more constrained than has previously been assumed.

What I want to do next is further motivate the claim that much of the workload of so-called polysemy can be re-allocated to either ambiguity or semantic indeterminacy. In section 5.2, I will illustrate some of what I consider genuine cases of polysemy and the generalisations they lead to.

5.1.3 Processing Considerations

One of the central tasks of a theory of polysemy is to give a psychologically tractable account of how a polysemic complex is actually represented, accessed and processed during on-line comprehension. In fact, the very definition of polysemy crucially depends on the form of the underlying lexical representation of a polysemous item: in order for polysemy to be recognised as a separate phenomenon, its underlying representation has to differ in significant respects from that of both monosemous and ambiguous items. For instance, the representation of a monosemous term such as *thermostat* is supposed to open up a unitary address in one's mental lexicon:

(23) *thermostat* -> THERMOSTAT

By contrast, ambiguous items such as *bank* and *port* correspond to more than one lexical entry in the mental lexicon:

(24) *bank* -> $BANK_1$ ('financial institution')
bank -> $BANK_2$ ('river edge')

(25) *port* -> $PORT_1$ ('wine')
port -> $PORT_2$ ('harbour')

In the second case, speakers may possess etymological information to the effect that both lexical addresses can originally be traced to a single lexical source. However, this information has the form of a redundancy rule and does not participate in the synchronic computation of the meaning of *port.* The question which arises now is: how is the mental representation of polysemy to differ from the possibilities in both (23) and (24)-(25)?

Different versions of current theorising about polysemy tacitly rely on different assumptions about the storage and subsequent processing of polysemous items. I will focus on cases where the range of different meanings of a given lexical item can be traced back to a single, more abstract schematic meaning: English modals fall into this category. On the most neutral interpretation of Sweetser's (1990) claims about polysemy, the representation of, say, *must* looks roughly like this:

(26) *must* -> FORCE-DYNAMIC COMPULSION - $MUST_1$ (root)
-> (metaphor) - $MUST_2$ (epistemic)

An overarching image-schematic concept, the force-dynamic concept of compulsion, organises the polysemic entry for *must.* When applied to the domain of physical experience, the result is a root modal meaning; when the image-schema projects into the abstract domain of reasoning, the epistemic meaning of *must* is produced. The priority of the root over the epistemic meaning is captured by the explicit representation of the metaphoric motivation of the sense extension into the epistemic domain.

I will call representations such as (26) *cluster models* (cf. Langacker's (1988) 'network models'). Cluster models allow for - although they do not require - the presence of an overarching general meaning. Where such a meaning exists, as in (26), the specific senses listed in the entry are instantiations of the blueprint, linked to each other through relations of similarity. Speakers of the language, the claim is, are sensitive to such relations and have complex intuitions about how the various sub-entries in polysemous items are synchronically related (Gibbs, 1994; Lakoff, 1987a; Taylor, 1995). Cluster models capture the difference between polysemy and either monosemy or ambiguity: polysemy, unlike ambiguity, is a property of unitary lexical entries; polysemic lexical entries, unlike monosemic ones, internally branch off to multiple interconnected sub-entries. In this respect, the representation in (26) occupies a middle ground between those in (23) and (24)-(25).

There are at least three ways of construing cluster models from the point of view of lexical comprehension. A first possibility is to take the various senses of a polysemous lexical item (i.e. the nodes of the cluster) to be individually memorised. Their conceptual relation will then be represented as a redundancy rule connecting two (or more) individual sub-entries within a single lexical address. During utterance comprehension, the nodes will be jointly activated and processed, while their conceptual links will be secondarily available as sources of intuitions about the relatedness of the separate senses. This option, which may be called the 'nodes only' approach, is compatible with a lot of writings on polysemy. However, it faces a number of problems. Firstly, if the synchronically-felt connection among the sub-parts of a single lexical entry is not recovered during utterance comprehension, then there is in practice very little to differentiate polysemy from ambiguity. Notice that, apart from cases like *bank*, where no connection is felt between the two meanings conveyed by the same form, there are ambiguous words such as *port* above for which some sort of connection exists and is recoverable by speakers of the language; the 'nodes only' account of polysemy conflates this type of ambiguity with classical cases of polysemy. This may not be objectionable in itself; nevertheless, it is a conclusion hard to reconcile with the wish to separate ambiguity from polysemy on principled grounds.

The second problem for the 'nodes only' solution relates to the first. If polysemy simply consists in the attachment of a redundancy rule (or a set of redundancy rules) to an internally complex lexical entry, then it is not a particularly interesting aspect of lexical semantics. Etymological links can be viewed as yet another type of lexical relation such as synonymy, hyponymy and so on, which operates throughout the mental lexicon. These relations do not play a role in on-line lexical understanding but may be used in post-access inference. Seen in this light, redundancy rules, although they may be considered as part of linguistic knowledge broadly construed, are not synchronically relevant for the mental representation of word meanings.

Finally, this view entails that the senses of a polysemous item are distinct, enumerable and mutually exclusive: this is most certainly not the case in the majority of examples. In the case of modality, as Coates (1983) has shown, it may be difficult to separate all the possible interpretations of a modal verb. The harder it is to systematically separate the contextual readings of a lexical item, the less reasonable it becomes to prefer a storage over a computation account of the existence of multiple interpretations.

A second option for psychological accounts of polysemy is to consider that the polysemy cluster is accessed exhaustively. On the 'full access' account, both the hyper-sense and the lower nodes are activated and processed during on-line comprehension. As a result of pragmatic processes of understanding, all nodes will be suppressed apart from the intended sense, together with its relation to the hyper-sense. Unlike the previous solution, the 'full access' proposal is capable of distinguishing polysemy from ambiguity on principled grounds, for the semantic relation among the senses of a polysemous item participates in utterance comprehension, unlike what happens in ambiguity processing. However, certain questions remain. To begin with, it is hard to motivate such a solution in terms of cognitive economy: why should the cognitive mechanism go through the computation of the various meanings in a polysemic network if the results are already streamlined (and accessible)? If no computation is involved and the conceptual relation motivating polysemy is simply added on top of the lower nodes, we are back to redundancy rules and their problems. Moreover, both the 'nodes only' and the 'full access' positions, by appealing to post-access suppression mechanisms, make the following prediction: the more polysemous a word is (i.e. the larger the underlying network of senses), the greater the extent of sense suppression required, hence the longer it should take to process. Experimental response times in verification tasks show that this prediction is not borne out (Grober, 1976; Caramazza and Grober, 1978).

There is a *prima facie* more plausible variant of the 'full access' approach: understanding polysemy does not involve simply retrieval but on-line derivation of the intended meaning. On this view, what is accessed is simply the superordinate sense and the mappings which will instantiate the more specific meanings of a polysemous item. Depending on contextual considerations, certain mappings will be suppressed and the correct one retained. In order to preserve the assumption that some meanings are more basic than others within the cluster, this model has to accommodate some hierarchy of mappings. In the example of *must* in (26), this postulates directionality from the root to the epistemic meaning, so that the latter cannot be accessed without the former. I will call this the 'hierarchical access' position.

The dependency of some meanings on others within a polysemy cluster is also reflected in the psycholinguistic literature, which often recognises degrees of dominance (or centrality) among the senses of a polysemous word. Durkin and Manning (1989) conducted a study in which subjects were asked to detect similarities between the meaning of a polysemous word embedded in an utterance, and a target word which was primed by either the dominant meaning or a non-dominant meaning of the polysemous word. Given a word like *fortune*, they established prior to the experiment that the word communicates two meanings which are judged as highly related. On the basis of production frequency, they rated the sense 'wealth' to be dominant and the sense 'good luck to be non-dominant. They then asked subjects to detect similarities between each of the occurrences of *fortune* in (27) on the one hand, and each of the two targets, *wealth* and *luck,* on the other:

(27) a. I won a fortune on Lotto.
b. I tested my fortune on Lotto.

They found that dominant meanings ('wealth') were more likely to be correlated with utterances biased towards the non-dominant meaning such as (27b) than vice-versa.

More interestingly, Williams (1992) employed a lexical decision task (familiar from psycholinguistic studies of ambiguity) for polysemous words using similar material. One of his examples was *dirty*, which can be interpreted as 'soiled' (dominant meaning) or 'obscene' (non-dominant meaning). He embedded the word in sentences priming one or the other interpretation and tested word recognition for the targets *soiled* and *obscene* in both types of sentences:

(28) a. The comedian was banned from television because his jokes were so dirty.
b. After playing football his legs were very dirty.

He found that recognition of targets related to dominant meanings in unrelated environments (e.g. *soiled* presented after (28a)) was primed regardless of delays, and that the priming did not diminish over time. However, priming of targets related to the non-dominant meanings in unrelated environments (e.g. *obscene* presented after (28b)) was weaker and tended to diminish over time. Moreover, these effects were associated with utterance comprehension, since isolated prime-target pairs (e.g. *dirty-soiled* vs. *dirty-obscene*) did not show any degree of differential facilitation. It would seem, therefore, that, quite apart from introspection, on-line processing data support the conclusion that some senses of polysemous items are privileged, or that a certain directionality exists in the way the various senses are produced from an over-arching blueprint.[8]

I think that some version of the 'hierarchical access' model has to be correct. That is, in order for a lexical item to qualify as polysemous rather than ambiguous, it must be that its non-central meanings are derived on-line from the central sense(s) through a specified process. Furthermore, in order for polysemy to differ from semantic indeterminacy, this process must be more constrained than fully inferential pragmatic work. So we have a way of making certain cases of polysemy intelligible, without reducing them to either ambiguity or monosemy. However, if one is to pursue this option within any of the current cognitive models of polysemy, one has to be more precise about both the general machinery used and the architecture of specific lexical entries. For instance, assuming that polysemy involves (in part) image-schematic mappings between or within conceptual

[8] A further, rarely considered possibility is that clusters are accessed differentially. This idea (which one might call the 'variable access' solution) is consistent with the network-like organisation of polysemy, but allows different parts of the entry to be accessed according to the demands of communication. Taylor (1995) puts forth this view in order to account for the variable acceptability of zeugmatic constructions with polysemous words. It is not clear what sort of constraints the 'variable access' position needs in order to determine which aspect of the cluster the interpretation process will focus on. I will not consider this option in what follows.

domains, we have to answer questions about how these mappings are fleshed out or how domains are individuated.

Moreover, in order to cash out the psycholinguistic findings about polysemy, one has to show that the (rather crude) psycholinguistic notion of dominance can be correlated with semantic basicness/centrality as recognised by theoretical analyses. In the remainder of this section, I want to argue that dominance in experimental work generally coincides with typicality, but the ascription of typicality values to the different senses of an (arguably) polysemous lexical item does not necessarily obey basicness ratings among the meanings of the item.

Consider the example of the adjective *long*, which has both a spatial and a temporal reading. It has been argued that 'the spatial sense is generally taken to be more central, *or prototypical*, and the temporal sense is related to it via metaphor' (Lakoff, 1987: 416-7, my emphasis). Thus, typicality judgements, at least in this case, are taken to recapitulate the (synchronically felt) dependency relation of the temporal upon the spatial sense. Here is the problem: if the directionality of metaphorical and other mappings within the lexicon accurately mirrors typicality ratings, then there is bound to be little disagreement among speakers as to the typicality of the senses of a polysemous word. Which sense is basic (and, hence, prototypical) is determined by the structure of the metaphor, and should thus be the same across individuals in a given community. Similarly, there should be little disagreement within an individual as to which sense is to instantiate the prototype for the use of the word.

It turns out that both of these predictions are erroneous. First, there is no obvious reason why this particular class of typicality judgements should prove to be immune from the discrepancies usually observed in prototype formation and evaluation across individuals, cultures, times and contexts. More importantly, there is some empirical evidence that in practice disagreements do exist. Durkin and Manning (1989) report that judgements of subjects as to the dominance (:basicness) of one sense of a polysemous item over another were largely a product of the sentence context. In the absence of a biasing context, 'the dominant meaning of a polysemous word is taken to be that which comes more readily to mind' (ibid. p. 589). In an already familiar study, Caramazza and Grober (1976) asked subjects to produce typicality judgements for various 'senses' of the word *line*, as these appeared in several sentences. They found that the highest typicality ratings were awarded to senses for which it had been established (independently of the experiment) that they had the highest frequency ratings. Both accessibility and frequency effects of the type demonstrated in the two experiments are generally assumed to have a bearing on the construction of prototypes (Barsalou, 1992); in particular, they have been used to explain the variation in typicality judgements across subjects, times and situations. However, these effects are not compatible with the position that subjects' typicality ratings were based on the derivational (semantic) priority of one lexical meaning of a polysemous word over others. Semantic basicness is presumably fixed (cf. the *long* example), while typicality judgements vary.

Second, there are examples of polysemous words where typicality judgements do not actually reflect basicness judgements. Consider the word *model*. For the sake of argument, I assume that it is polysemous and that its senses may roughly be paraphrased as follows: (i) 'reproduction/representation of something', (ii) 'example to imitate', (iii) 'person who poses for artists', (iv) 'mannequin' (I simplify for exposition). One may plausibly assume that the first listed sense is the basic one, from which the other senses are derived via metaphoric and other processes (this would conform with the historical facts, so that polysemy recaptures internal semantic reconstruction - Traugott, 1986). However, I strongly doubt that this 'basic' sense would fare very well on the typicality ratings; in fact, a preliminary investigation with native speakers shows that (ii) and (iv) are more plausible candidates. Obviously, a lot more needs to be said about this example. If generalised, however, similar cases may provide counterexamples to the connection between prototypes and semantic basicness.

The above discussion has suggested that, despite the confusing assumptions surrounding this notion, there is a way of making polysemy consistent and psycholinguistically intelligible. As a result, however, the scope of this phenomenon is greatly reduced. In the next section, I want to make some suggestions as to how both apparent and genuine cases of polysemy may be handled within a more inclusive picture of the psychology of word meaning.

5.2 Inference and the Plasticity of Lexical Concepts

5.2.1 A Reanalysis of Certain Polysemy Phenomena

So far, I have claimed that a number of factors which have been assumed to require a polysemy analysis for a large class of lexical items can, in fact, be accommodated by either ambiguity or semantic indeterminacy of some sort. This opens up the way to a reanalysis of many individual cases. To take one well-known example, *open* appears to be able to convey a large number of interconnected meanings in different contexts:

(29) a. Open the office/a parcel/an envelope.
b. Open the door/the lid/the cork.
c. Open one's shirt/a zip/a penknife.
d. Open an exhibition/a nature reserve/a road.
e. Open a discussion/a debate/a conference.

On a polysemy analysis, *open* encodes a cluster of concepts which correspond to the separate meanings in (29a-e) and which are brought together by criss-crossing levels of similarities. (29a) involves the idea of gaining access to the interior of a closed container; (29b) focuses on the manner a device may be manipulated so as to create an aperture; the use in (29c) focuses on the moving apart of the component parts of an entity; (29d) has to do with making something accessible to the general public; finally, (29e) refers to the initiation of an interactional process (Taylor, 1995: 287).

Notice firstly that not all of the above meanings would pass the classic ambiguity tests. In particular, there seems to be a distance between (a)-(c) on the one hand and (d)-(e) on the other, so that the utterances in (30) feel increasingly zeugmatic:

(30) a. (?)I asked him to open the door and a parcel.
b. (?)I asked him to open the lid and a bottle of wine.
c. ??I asked him to open the envelope and the exhibition.
d. *I asked him to open the invitation and the debate.

On the basis of similar results, one may propose that *open* is not a unitary lexical entry but is ambiguous and corresponds to two lexical items: *open*$_1$ might include the meanings in (a)-(c) and *open*$_2$ those in (d)-(e). Within each of these items, *open* can be thought to be univocal: depending on the sort of thing one applies the predicate to (typically the object of the verb), *open* communicates a slightly different concept. Following Sperber and Wilson (1986/1995, 1997), I assume that the precise concept conveyed by this expression will be shaped by considerations of relevance. It is reasonable to suppose that (29c)/(29d) were previously creative, pragmatically derived interpretations of *open*$_1$; after intermediate degrees of lexicalisation, they formed a stable and independent lexical entry. Although the etymological connection (and hence the conceptual relation) with the parent concept are synchronically retrievable, they play no role during utterance comprehension.

The case of *good* is not dissimilar:[9]

[9] For a polysemy analysis of *good*, see Pustejovsky (1995).

(31) a. good weather
b. a good book
c. a good student
d. good food

The difference from *open* (ambiguity aside) is that *good* is semantically incomplete: being a relational adjective, it requires contextual specification of an 'unset parameter' in its lexical meaning (corresponding to the measure of goodness). So far I have used 'semantic indeterminacy' to cover both generality of meaning (the *open*$_I$ case) and semantic incompleteness (the *good* case): my main aim was to show how either version of indeterminacy could deal with apparent cases of polysemy (for the distinction, cf. chapter 1). Recall that a similar situation arose with modality: in chapter 2, I motivated two alternatives to polysemy analyses of the English modals, one involving free contextual enrichment for semantically complete entries (i.e. entries with no gaps in their semantic representation: e.g. *can*), the second involving contextual saturation for semantically incomplete entries (i.e. entries with a slot in their semantic representation which needs to be pragmatically filled in: e.g. *must*).

As the previous examples already demonstrate, the main bulk of the reanalysis of polysemy falls on the side of pragmatics. This shift in perspective has the advantage of avoiding the proliferation of word senses which led to intricate usage-based models with few internal constraints. For instance, according to the rationale of the polysemy analysis of *open*$_I$, the discovery of new ways of opening things would bring about a change in the meaning of *open* (a local reorganisation of the polysemy cluster it encodes). The reliance on semantics to provide stored representations of a large number of contextually attested interpretations goes a considerable way back towards the code model of communication.[10] On this model, communication is simply a matter of successfully encoding and decoding messages into and from appropriate signals; inference plays a minimal role, and where it appears, it works in a restricted way circumscribed by code-like instructions. Many current models adopting a semantic maximalism (polysemy models included), although very different in persuasion from code models in other respects, share with them a distrust of purely inference-driven accounts of the context-dependence of lexical items.

Far from being arbitrary and unconstrained, however, a full-fledged inferential model of how lexical meaning is constructed in context is, in fact, supported by much recent research on concept manipulation in communication. Works such as Barsalou (1982, 1987), Braisby and Franks (1990), Franks (1995), Goshke and Koppelberg (1992), Butler

[10] See the discussion in Sperber and Wilson (1997), Wilson (1998).

(1995) demonstrate from alternative perspectives the ubiquity of the ad hoc formation of concepts from underlying representations in various contextual conditions. From a relevance-theoretic perspective, it is to be expected that we have many more concepts than natural language words, and hence that we rely on powerful inferential mechanisms to derive a pragmatic instantiation of what may be a rather different semantically encoded content. On this account, natural-language words are simply pointers to the concept which the speaker intends to communicate in a given situation: cost-effect processing considerations, the hearer's expectations and the evaluation of the speaker's intentions all enter into the computation of the concept which is expressed (though not necessarily encoded) by a given linguistic item.

Certain aspects of this inferential process may become stabilised after repeated access. Various types of typicality effects may be explained in this model as by-products of the frequency and accessibility of certain conceptual attributes over others within a single conceptual address. Neither does a pragmatic account preclude the possibility that some word uses are stored as pre-formed units and accessed from memory rather than being constructed *ab ovo*: a case in point might be the previous example *line of reasoning*, for which it is reasonable to assume that it has become a separately encoded construction. The disagreement between a fully inferential and a usage-based model rests on the proportion of concepts for which separate storage and activation is recognised.

Monosemy-based accounts of word meaning can also accommodate differences among speakers in the way lexical concepts are stored and understood. What may be an ambiguous term for one speaker may be understood fully pragmatically by another; what may be a stable, lexicalised concept for one member of the linguistic community may be a creatively produced and/or comprehended concept for another. Frequency and accessibility of use and the size and accessibility of context which prompts a certain interpretation influence the extent and ease of lexicalisation of what were previously one-off uses of lexically encoded concepts.

5.2.2. Some Residual Cases

I have argued at length for a complex interplay between semantic and pragmatic factors in detrmining what is expressed by a lexical item in a specific situation of utterance. Moreover, I have claimed that the concept of polysemy has a more restricted role to play in this process than is sometimes assumed. Nevertheless, both in syntactic tests and in experimental results on on-line processing there has been an indication that there are certain phenomena which cannot properly be reanalysed as aspects of either pragmatic manipulation or ambiguity. I now want to turn briefly to this residue of cases, which exhibit systematic meaning alternations and qualify for genuine polysemy. Consider the following examples:

(32) book copy/content
a. This book is really heavy.
b. This book is really difficult.

(33) container/content
a. The glass broke.
b. She's had three glasses in an hour.

(34) newspaper copy/institution
a. Did you read the newspaper today?
b. John works for a big newspaper.

These and other pairs have had some notoriety in recent writings on lexical semantics. It appears that they correspond to generally productive alternations within the lexicon, although some of their outputs are lexicalised to a greater extent than others and productivity is overall curtailed by pragmatic considerations and encyclopedic knowledge. These examples would pass syntactic tests as in (35), which suggests that they cannot properly be relegated to the case of ambiguity:

(35) *The Press*, which is owned by one of the biggest conglomerates in the country, usually has only local news.

On the other hand, the relative predictability of these shifts cannot be captured by a fully creative pragmatic account. One way of handling these alternations is to consider them as semi-automated pragmatic routines which take as input one member of the pair and give as output the other member. Although in principle their input ranges over whatever satisfies their structural description, in practice the acceptability of particular outputs depends on pragmatic factors. As a result, different cultures/communities sanction different routines. For instance, in Modern Greek it is possible to use the name of a part of the body to denote an illness which affects that part. In English, however, *Jane has a/her heart* cannot mean that Jane suffers from heart problems. Such non-linguistic conventions/regularities have a local application and are subject to constraints such as blocking or pre-emption from already lexicalised cases (see example (16) above).

Even within this more constrained range of application, though, polysemy is not a unified phenomenon. Rather than treating all systematic alternations of the above type as pragmatic licences, for at least some cases one needs a stronger, semantic treatment. There are kinds of meaning multiplicity which, although superficially very similar to (32)-(34), are accompanied by grammatical (phonological, morphological or syntactic) constraints. Examples include the use of the name of a fruit/nut to denote the tree of the fruit/nut in Spanish and Italian, which is accompanied by a change of gender (masculine for the tree; see Copestake and Briscoe, 1996: 17). Similar cases are often the province of zero-derivation morphological rules. In English, there is a rule which creates denominal verbs such as *butter*, *hammer*, etc. (Clark and Clark, 1979; Copestake and Briscoe, 1996): although its conceptual motivation may lie with pragmatic alternations such as (32)-(34), the rule is part of the grammar of the language.[11] Since these rules result in the creation of a novel lexical item, they should not be considered as proper polysemy.

I cannot elaborate on these examples further. The conclusion from this discussion appears to be that some genuine cases of polysemy can be recognised and placed between radical semantic and radical pragmatic phenomena. These cases involve systematic derivation rather than retrieval of lexical meaning (a prerequisite for genuine polysemy, as we saw). The derivation may range from pragmatic, though local/streamlined processes or routines to grammatical rules; its further end may involve word-formation rules which share with polysemy a general conceptual motivation, although they result in different lexical items.

5.3 Historical and Developmental Perspectives

Polysemy analyses make regular reference to historical and developmental arguments to motivate a cluster-based rather than a monosemy or ambiguity account. The claim is that neither ambiguity nor 'flat' monosemy can capture regularities in diachronic and developmental meaning changes, which can be naturally represented within polysemy networks. This claim can be unpacked into two, often interconnected, assumptions. Firstly, it is held that each of these domains will provide crucial and relevant evidence for the shape of synchronic adult lexical knowledge. Secondly, it is expected that the

[11] For further examples and discussion, see Pelletier and Schubert (1989), Levin (1993), Pustejovsky (1993, 1995), Hale and Keyser (1993), Lieber (1980).

directionalities in both domains will (at least frequently) non-accidentally coincide and may also reflect the judgements of competent speakers about what they know about the meanings of the words in their language.

Let me start with one aspect of the first claim, i.e. the bearing of developmental arguments on adult lexical competence. At first sight, the connection is open to question. Notice that it is generally accepted that the child's lexicon falls short of an adult's; moreover, the development of specific aspects of the child's capacities is normally traced against what we know about the full-fledged linguistic capacities of adults. It would then seem peculiar to use children's competence the other way around as well, i.e. to take it as indicating something about the organisation of adult competence. Acquisitional data cannot work both ways, as evidence about both the starting point and the endpoint of semantic competence.[12]

The second claim can best be summarised as the 'ontogeny recapitulates phylogeny' tenet. Stephany (1979/1986), among others, has explicitly put forth this view with regard to modality in maintaining that developmental progress re-iterates diachronic grammaticalisation processes. She concludes her article by stating: 'The priority of deontic, as compared to epistemic, modality in the ontogenesis as well as in the history of languages can be considered as indicating the primacy of the social, as compared to the epistemic, function of language' (ibid. p. 400).

It turns out that historical change and developmental progress do not always move on parallel paths. As the Appendix 3A has shown, the 'permission' reading of *may* appears historically later than the epistemic reading, and both are derived from an original 'ability' reading of the verb. By contrast, in child language the epistemic use of the modal lags behind its deontic use. Such examples could be multiplied. It is worth mentioning, for instance, that the set of the English modal auxiliaries historically developed from a set of main verbs; this development has no parallel in acquisition.

The main reason for considering historical and developmental evidence as synchronically relevant is the presence in both cases of intermediate stages of lexical representations. In child development, it is claimed that these intermediate stages before the adult-like lexical entry is constructed remain stored even if the final adult entry includes a core meaning. Similarly, in historical change of lexical meaning, previous linguistic states are synchronically available to the speakers and contribute to cluster representations of word meaning. So failure to erase previous lexical-semantic representations creates polysemy (see Taylor, 1995: 286).

---

[12] Of course, as Neil Smith pointed out to me, the developmental progression the child goes through may offer various insights not provided by either its initial or its steady states.

It is undeniably true that speakers of a language store a great deal of information about lexical items: etymological information and various other connections among lexical entries or sub-entries (synonyms, hyponyms, and so on) belong to this kind of information. However, as I have mentioned earlier in this chapter, this information does not (at least necessarily) participate in the lexical comprehension process but may serve as the starting point for many post-comprehension lexically-driven inferences. The re-organisation of lexical entries during ontogenetic and phylogenetic progress does leave behind all sorts of information which has somehow been made redundant or outdated by recent developments. In cases where this information is not erased (and this is certainly possible), it remains available somewhere in the lexical entry, but does not influence what is perceived as the (synchronic, adult-like) meaning of the item. Hence arguments which show the internal richness of lexical *information* do not immediately translate into arguments for the internal complexity of lexical *meaning*.[13]

The conclusion seems to be that the study of the directionality and succession of stages of lexical development cannot yield direct insight into the organisation of the adult lexicon. Rather, developmental models and accounts of adult semantic competence merely have to be compatible with each other and mutually informed: in the case-study of modality, I have tried to use the metarepresentation hypothesis as a means of ensuring that the proposed semantics and pragmatics for modality make developmental sense. Likewise, accounts of diachronic change do not preclude the possibility that speakers keep track of meaning changes across time, even though they access and use synchronically univocal semantic entries.

5.4 Concluding Remarks

My aim in this chapter was to reconsider certain assumptions about word meaning and conceptual structure paying special attention to the notion of polysemy. I have argued that polysemy in natural language is not a natural class. More specifically, I have claimed that many of the phenomena it subsumes can be reanalysed as either monosemy or ambiguity. I have also attempted to counter a broad consensus in the literature on lexical semantics, according to which these alternative semantic options cannot capture basic facts about how word meaning is stored, understood, acquired and modified across time.

---

[13] *Mutatis mutandis*, the same can be held for ambiguity analyses.

This is not to say that monosemy is the null hypothesis, or that all of our lexical concepts should have uniform (and unitary) semantic representations. On the contrary, this research suggests that both the form of underlying lexical semantic representations and the way they interact with the pragmatic device are complex and varied. Nevertheless, polysemic structures are not always a useful way of representing the flexibility and context-dependence of lexically expressed meaning. If the arguments of the previous pages are correct, alternative analyses based on inferential pragmatic processes can successfully deal with large part of the systematic variability of lexical concepts in communication.

6

CONCLUSION

The main body of this book was taken up by a detailed analysis of the semantics and pragmatics of a sample of the English modal verbs. I have argued for a unitary semantic treatment of *may*, *must*, *can*, *should* and *ought to*, and have given a systematic exposition of the ways in which different interpretations for these verbs arise during on-line communication. I have adopted the view that modal expressions correspond to tripartite quantificational structures which admit different contextually supplied domains of quantification. In this sense, my analysis builds on previous formal approaches that have explicitly recognised the role of contextual restrictions on quantificational domains. A crucial difference from previous accounts of modality is the proposal to treat modal verbs in their epistemic interpretations as metarepresentational operators. On this view, epistemic modals mark operations on mental representations of propositions entertained *as mental objects*, and hence reflect basic metacognitive ('theory-of-mind') abilities. Using a relevance-theoretic framework for the analysis of communication, I have shown how the ability to metarepresent, together with a variety of pragmatic considerations, can predict the range of interpretations English modal verbs receive in context.

My approach departs from previous (mainly polysemy-based analyses) on a number of further points. Firstly, I have presented some reasons for doubting that the distribution of root and epistemic interpretations is generally predictable on the basis of the configurational properties of the modals: it is preferable to view the alleged grammatical reflexes of the root-epistemic distinction as natural aspects of the interpretation process. Secondly, I have reconsidered the priority of root over epistemic meanings in language acquisition, which has long been associated in the psycholinguistic literature with polysemy in modal expressions. I have argued that the link between epistemic modality and metarepresentation places a threshold on the development of epistemic modal expressions (applied in conjunction with several other factors), without threatening in any way the claim about the unitary meaning of the modals. Thirdly, I have demonstrated that the mixed evidence for the truth-conditionality of epistemic modality falls out from the metarepresentational properties of epistemic complements and relevance considerations, and does not provide an argument for semantically separating root from epistemic modal meanings. Fourthly, I have claimed that the category of speech-act modality (accepted by some researchers as a separate type of modality) can be analysed away as a distinct sub-type of the metarepresentational use of language. Finally, I have considered the claim that root modal expressions in various languages come to express epistemic concepts at a later historical stage. Consistently with my overall analysis, I have not taken this fact to indicate polysemy in modals but made some tentative suggestions of a link between subjectification in linguistic change and the development of metarepresentational readings in modals.

I started this book by asking: what is it about the semantic representation of lexical items that allows them to receive multiple interpretations in context? The case-study of modality has motivated two types of starting points provided by grammatical information: semantic generality (which may bring about some form of free enrichment, as in *can*, *should* and *ought to*), and semantic incompleteness (which requires pragmatically induced saturation, as in *must* and *may*). Both these types are instances of semantic underdeterminacy: the information which the modals semantically encode forms simply the input to pragmatic mechanisms of utterance comprehension, which are powerful enough to yield the contextual variability of modal interpretations. Semantic underdeterminacy of this kind starkly contrasts with the semantic overdeterminacy adopted by polysemy-based explanations of the contextual flexibility of lexical content.

Apart from the reanalysis of arguments pertaining specifically to modal phenomena, I have claimed that a range of independent factors which have been taken to support polysemy analyses (e.g. the intuitive similarity between different readings of a lexical item, the cross-linguistic translatability of contextually attested interpretations, etc.) are inconclusive. It appears, then, that polysemy is neither tightly defined nor overwhelmingly evidenced by the arguments which are commonly cited in its support; and, although there are some aspects of the phenomenon which seem to be sufficiently circumscribed and cognitively real, their scope is more constrained than previous semantic approaches have claimed. This leaves ample space for dealing with much of the plasticity of lexically expressed information in pragmatic terms. As one might expect on an inferential account of communication, the concept encoded by a lexical item is simply a starting point for the construction of the contextually appropriate (:intended) construal of the concept.

This conclusion leaves open various possibilities for further research. On the empirical side, an obvious supplement to the work done in this book would be an extension of the proposed framework to the rest of the modal verbs (and other modal expressions) in English. Moreover, a more detailed investigation based on a wider array of cross-linguistic modal data would be very useful. On a different note, a number of themes have emerged in the course of the discussion in the previous chapters, which raise some interesting theoretical questions. Here I will only mention four.

In the first place, I have only outlined some of the issues surrounding the mental representation of word meaning. Further research is needed to clarify the mechanisms whereby lexical concepts systematically rely on aspects of context in communication. Moreover, the territory of genuine polysemy remains to be charted with precision. For instance, it seems that polysemy appears in local alternations of the sort 'container/content' which I mentioned briefly towards the end of chapter 5. One would like to know more about the way in which these alternations work, how they relate to morphosyntactic rules (e.g. word-formation rules), and how their semi-productivity is to be represented in the grammar.

Secondly, the concept of metarepresentation seems a promising area for much interdisciplinary work. Obviously, the sort of analysis which I have proposed for epistemicity and evidentiality can be extended to further linguistic areas, since several aspects of semantic and pragmatic development seem to presuppose theory-of-mind understanding. Some recent directions within this newly-set research agenda range from reference assignment and conversational skills to the early appreciation of intention in word learning (see de Villiers and Fitneva, 1998; Leslie and Happé, 1989; Bloom, 1995, 2000). Research on theory of mind establishes an exciting link between early conceptual and linguistic abilities, and can be used as an informative case-study in assessing competing accounts of the nature and development of mental architecture.

Thirdly, it remains to be seen how the notion of the main relevance of the utterance correlates with issues of truth-conditionality. From a relevance-theoretic perspective, the 'main point' of the utterance is naturally the locus of the cognitive effects which the utterance (intentionally) creates for the hearer, and intuitions about main relevance are expected to cross-cut (and be stronger than) intuitions about truth-conditional content. There seems to be a convergence on this point from other directions within recent relevance-theoretic research (see, e.g., Carston, 1998; Ifantidou, 1994; Blakemore, 1990/1; cf. Sperber and Wilson, 1984), so this is an issue which merits further attention.

Finally, a theme which indirectly arises from the previous pages relates to the boundaries and compatibility of formal and cognitive approaches to semantics. As I mentioned in the Introduction, one may maintain a distinction between 'real' (truth-theoretic) and linguistic (translational) semantics, and regard the two as separate enterprises with conceivably diverse domains of inquiry and methodological constraints. Still, considerations of psychological plausibility are increasingly introduced into formal semantic accounts (see Partee, 1995b). In the case of modality, my analysis was articulated with the purpose of providing a psychologically informed version of previous formal approaches to modal categories in natural language. It is a challenge for the next stages of semantic theory to combine formal elegance and psychological reality in a single research program, and develop into an integrated and cognitively robust field of linguistic inquiry.

References

Adger, D. (1997). Back to modals: some cross-linguistic generalisations. Talk delivered at the Department of Linguistics, University College London, Fall.

Aksu-Koç, A. (1986). *The Acquisition of Aspect and Modality: The Case of Past Reference in Turkish.* Cambridge University Press, Cambridge.

Aksu-Koç, A. and D. Slobin (1986). A psychological account of the development and use of evidentials in Turkish. In: *Evidentiality: The Linguistic Coding of Epistemology* (W. Chafe and J. Nichols, eds.), pp. 159-167. Ablex, Norwood, NJ.

Anderson, L. (1986). Evidentials, paths of change and mental maps: typologically regular asymmetries. In: *Evidentiality: The Linguistic Coding of Epistemology* (W. Chafe and J. Nichols, eds.), pp. 273-312. Ablex, Norwood, NJ.

Anscombe, G. (1957). *Intention.* Oxford University Press, Oxford.

Apresjan, J. (1973). *Regular Polysemy.* Mouton, The Hague.

Astington, J. (1990). Wishes and plans: Children's understanding of intentional causation. Paper presented at the Annual Meeting of the Jean Piaget Society, Philadelphia, PA, May.

Astington, J. (1991). Intention in the child's theory of mind. In: *Children's Theories of Mind* (D. Frye and C. Moore, eds.), pp. 157-172. Erlbaum, Hillsdale, NJ.

Astington, J. and A. Gopnik (1991). Theoretical explanations of children's understanding of the mind. In: *Perspectives on the Child's Theory of Mind* (G. Butterworth, P. Harris, A. Leslie and H. Wellman, eds.), pp. 7-31. Oxford University Press/The British Psychological Society, Oxford.

Atlas, J. (1989). *Philosophy without Ambiguity.* Clarendon Press, Oxford.

Bach, K. (1994a). Semantic slack: What is said and more. In: *Foundations of Speech Act Theory: Philosophical and Linguistic Perspectives* (S. Tsohatzidis, ed.), pp. 267-291. Routledge, London and New York.

Bach, K. (1994b). Conversational impliciture. *Mind and Language*, **9**, 124-162.

Bach, K. and R. Harnish (1979). *Linguistic Communication and Speech Acts.* MIT Press, Cambridge, MA.

Baldwin, D. (1991). Infants' contribution to the achievement of joint reference. *Child Development*, **62**, 875-890.

Baldwin, D. (1993). Infants' ability to consult the speaker for clues to word reference. *Journal of Child Language*, **20**, 395-418.

Barbiers, S. (1995). *The Syntax of Interpretation*. HIL Dissertation Series, Leiden University.

Baron-Cohen, S. (1989). The autistic child's theory of mind: A case for specific developmental delay. *Journal of Child Psychology and Psychiatry*, **30**, 285-297.

Baron-Cohen, S. (1995). *Mindblindness: An Essay on Autism and Theory of Mind*. MIT Press, Cambridge, MA.

Baron-Cohen, S., A. Leslie and U. Frith (1985). Does the autistic child have a 'theory of mind'? *Cognition*, **21**, 37-46.

Baron-Cohen, S., H. Tager-Flusberg and D. Cohen, eds. (1993). *Understanding Other Minds: Perspectives from Autism*. Oxford University Press, Oxford.

Barsalou, L. (1983). Ad hoc categories. *Memory and Cognition*, **11**, 211-227.

Barsalou, L. (1987). The instability of graded structure in concepts. In: *Concepts and Conceptual Development: Ecological and Intellectual Factors in Categorization* (U. Neisser, ed.), pp. 101-140. Cambridge University Press, New York.

Barsalou, L. (1992). *Cognitive Psychology: An Overview for Cognitive Scientists*. Erlbaum, Hillsdale, NJ.

Barsalou, L. (1993). Flexibility, structure, and linguistic vagary in concepts: Manifestations of a compositional system of perceptual symbols. In A. Collins, S. Gathercole, M. Conway and P. Morris (eds.). *Theories of Memory*, 29-89. Erlbaum, Hillsdale, NJ.

Bassano, D. (1996). Functional and formal constraints on the emergence of epistemic modality: A longitudinal study on French. *First Language*, **16**, 77-113.

Bierwisch, M. and R. Schreuder (1992). From concepts to lexical items. *Cognition*, **42**, 23-60.

Blakemore, D. (1987). *Semantic Constraints on Relevance*. Blackwell, Oxford.

Blakemore, D. (1988). 'So' as a constraint on relevance. In: *Mental Representations: The Interface between Language and Reality* (R. Kempson, ed.), pp. 183-195. Cambridge University Press, Cambridge.

Blakemore, D. (1989). Denial and contrast: A relevance theoretic analysis of *but*. *Linguistics and Philosophy*, **12**, 15-38.

Blakemore, D. (1990/1). Performatives and parentheticals. *Proceedings of the Aristotelian Society*, Vol. LXXXXI, 197-213.

Blakemore, D. (1992). *Understanding Utterances*. Blackwell, Oxford.

Blakemore, D. (1994). Evidence and modality. In: *The Encyclopedia of Language and Linguistics* (R. Asher, ed.). Pergamon Press, Oxford.

Blass, R. (1990). *Relevance Relations in Discourse: A Study with Special Reference to Sissala*. Cambridge University Press, Cambridge.

Bloom, P. (1995). Theories of word learning: Rationalist alternatives to associationism. In: *Handbook of Child Language Acquisition* (W. Ritchie and T. Bhatia, eds.), pp. 249-278. Academic Press, San Diego and London.

Bloom, P. (2000). *How Children Learn the Meanings of Words*. MIT Press, Cambridge, MA.

Bolinger, D. L. (1989). Extrinsic possibility and intrinsic potentiality: 7 on MAY and CAN +1. *Journal of Pragmatics*, **13**, 1-23.

Boyd, J. and J. Thorne (1969). The semantics of modal verbs. *Journal of Linguistics*, **5**, 57-74.

Bradley, R. and N. Swartz (1979). *Possible Worlds: An Introduction to Logic and its Philosophy*. Hackett Publishing Co., Indianapolis, Indiana.

Braine, M. and B. Rumain (1983). Logical reasoning. In: *Handbook of Child Psychology* (P. Mussen, ed.), Vol. III: Cognitive Development, pp. 263-340. John Wiley, New York.

Breheny, R. (1999). *Procedural Semantics*. Doctoral dissertation, University of London. To appear from Oxford University Press.

Brennan, V. (1993). *Root and Epistemic Modal Auxiliary Verbs*. Doctoral Dissertation, University of Massachusetts at Amherst.

Bretherton, I. and M. Beeghly (1982). Talking about internal states: The acquisition of an explicit theory of mind. *Developmental Psychology*, **6**, 906-921.

Brown, G. and G. Yule (1983). *Discourse Analysis*. Cambridge University Press, Cambridge.

Brown, K. (1991). Double modals in Hawick Scots. In: *Dialects of English* (P. Trudgill and J. Chambers, eds.), pp. 74-103. Longman, London.

Butler, K. (1995). Content, context, and compositionality. *Mind and Language*, **10**, 3-24.

Bybee, J. (1988a). Semantic substance vs. contrast in the development of grammatical meaning. In: *Proceedings of the 14th Annual Meeting of the BLS* (S. Axmaker, A. Jaisser and H. Singmaster, eds.), pp. 247-279. BLS, Berkeley, CA.

Bybee, J. (1988b). The diachronic dimension in explanation. In J. A. Hawkins (ed.). *Explaining Language Universals*, 350-379. Blackwell, Oxford.

Bybee, J. and S. Fleischman (1995a). Modality in grammar and discourse: An introductory essay. In: *Modality in Grammar and Discourse* (J. Bybee and S. Fleischman, eds.), pp. 1-14. Benjamins, Amsterdam.

Bybee, J. and S. Fleischman, eds. (1995b). *Modality in Grammar and Discourse.* Benjamins, Amsterdam.

Bybee, J. and W. Pagliuca (1985). Cross-linguistic comparison and the development of grammatical meaning. In: *Historical Semantics and Historical Word-Formation* (J. Fisiak, ed.), pp. 59-84. Mouton, Berlin.

Bybee, J., R. Perkins and W. Pagliuca (1994). *The Evolution of Grammar: Tense, Aspect, and Modality in the Languages of the World.* University of Chicago Press, Chicago and London.

Byrnes, J. and H. Beilin (1991). The cognitive basis of uncertainty. *Human Development*, **34**, 189-203.

Byrnes, J. and M. Duff (1989). Young children's comprehension of modal expressions. *Cognitive Development*, **4**, 369-387.

Byrnes, J. and W. Overton (1986). Reasoning about certainty and uncertainty in concrete, causal and propositional contexts. *Developmental Psychology*, **22**, 793-799.

Caramazza, A. and E. Grober (1976). Polysemy and the structure of the subjective lexicon. In: *Semantics: Theory and Application. Georgetown Round Table on Language and Linguistics* (C. Rameh, ed.), pp. 181-206. Georgetown University Press, Washington, DC.

Carruthers, P. and P. Smith, eds. (1996). *Theories of Theories of Mind.* Cambridge University Press, Cambridge.

Carston, R. (1988). Implicature, explicature, and truth-theoretic semantics. In: *Mental Representations: The Interface between Language and Reality* (R. Kempson, ed.), pp. 155-181. Cambridge University Press, Cambridge. Reprinted in: *Pragmatics: A Reader* (S. Davis, ed., 1991), pp. 33-51. Oxford University Press, Oxford.

Carston, R. (1993). Conjunction, explanation, and relevance. *Lingua*, **90**, 27-48.

Carston, R. (1994). Metalinguistic negation and echoic use. *UCL Working Papers in Linguistics*, **6**, 321-339.

Carston, R. (1996). Enrichment and loosening: Complementary processes in deriving the proposition expressed? *UCL Working Papers in Linguistics*, **8**, 205-232.

Carston, R. (1997). Informativeness, relevance and scalar implicature. In: *Relevance Theory: Applications and Implications* (R. Carston and S. Uchida, eds.), pp. 179-236. Benjamins, Amsterdam.

Carston, R. (1998). *Pragmatics and the Explicit-Implicit Distinction.* Doctoral dissertation, University of London. To appear from Blackwell as *Thoughts and Utterances.*

Carston, R. and E. Noh (1995). A truth-functional account of metalinguistic negation, with evidence from Korean. *UCL Working Papers in Linguistics*, **7**, 1-26.

Charman, T. and S. Baron-Cohen (1992). Understanding beliefs and drawings: A further test of the metarepresentation theory of autism. *Journal of Child Psychology and Psychiatry*, **33**, 1105-1112.

Choi, S. (1995). The development of epistemic sentence-ending modal forms and functions in Korean children. In: *Modality in Grammar and Discourse* (J. Bybee and S. Fleischman, eds.), pp. 165-204. Benjamins, Amsterdam.

Chung, S. and A. Timberlake (1985). Tense, aspect, and mood. In: *Language Typology and Syntactic Description* (T. Shopen, ed.), Vol. III, pp. 202-258. Cambridge University Press, Cambridge.

Cinque, G. (1999). *Adverbs and Functional Heads: A Cross-Linguistic Perspective*. Oxford University Press, Oxford.

Clark, E. (1993). *The Lexicon in Acquisition*. Cambridge University Press, Cambridge.

Clark, E. and H. Clark (1979). When nouns surface as verbs. *Language*, **55**, 767-811.

Clark, W. (1991). *Relevance Theory and the Semantics of Non-Declaratives*. Doctoral dissertation, University of London.

Coates, J. (1983). *The Semantics of the Modal Auxiliaries*. Croom Helm, London and Canberra.

Coates, J. (1988). The acquisition of the meaning of modality in children aged eight and twelve. *Journal of Child Language*, **15**, 425-434.

Coates, J. (1995). The expression of root and epistemic possibility in English. In: *Modality in Grammar and Discourse* (J. Bybee and S. Fleischman, eds.), pp. 55-66. Benjamins, Amsterdam.

Colombo, L. and G. Flores d' Arcais (1984). The meaning of Dutch prepositions: A psycholinguistic study of polysemy. *Linguistics*, **22**, 51-98.

Copestake, A. and T. Briscoe (1996). Semi-productive polysemy and sense extension. In: *Lexical Semantics: The Problem of Polysemy* (J. Pustejovsky and B. Boguraev, eds.), pp. 15-67. Clarendon Press, Oxford.

Cormack, A. and N. Smith (forth.). Modals and negation in English. In: *Modality in Generative Grammar* (F. Beukema and S. Barbiers, eds.). Benjamins, Amsterdam.

Cresswell, M. (1973). *Logics and Languages*. Methuen, London.

Cross, C. (1986). 'Can' and the logic of ability. *Philosophical Studies*, **50**, 53-64.

Cruse, A. (1986). *Lexical Semantics*. Cambridge University Press, Cambridge.

Cruse, A. (1995). Polysemy and related phenomena from a cognitive linguistic viewpoint. In: *Computational Lexical Semantics* (P. Saint-Dizier and E. Viegas, eds.), pp. 33-49. Cambridge University Press, New York.

Dancy J. (1985). *An Introduction to Contemporary Epistemology*. Blackwell, Oxford.

Dawson G., ed. (1989). *Autism: Nature, Diagnosis, and Treatment.* New York: Guildford Press.

de Roeck, A. and J. Nuyts (1994). *Epistemic Modal Expressions by High-Functioning Autistic Adults.* Antwerp Papers in Linguistics **82**.

de Villiers, J. and S. Fitneva. (1998). Referential transparency for opaque containers. Smith College ms./Under revision.

Dowty, D. (1986). The effects of aspectual class on the temporal structure of discourse: Semantics or pragmatics? *Linguistics and Philosophy*, **9**, 37-61.

Durkin, K. and J. Manning (1989). Polysemy and the subjective lexicon. *Journal of Psycholinguistic Research*, **18**, 577-612.

Ehrman, M. (1966). *The Meanings of the Modals in Present-Day American English.* Mouton, The Hague.

Eisenmajer, J. and M. Prior (1991). Cognitive linguistic correlates of 'theory of mind' abilty in autistic children. *British Journal of Developmental Psychology*, **9**, 351-364.

Espinal, T. (1991). The representation of disjunct constituents. *Language*, **67**, 726-762.

Fabricius, W., C. Sophian and H. Wellman (1987). Young children's sensibility to logical necessity in their inferential search behaviour. *Child Development*, **58**, 409-423.

Farkas, D. (1981). *Intensionality and Romance Subjunctive Relatives.* Doctoral dissertation, University of Chicago. IULC, Bloomington.

Farkas, D. and Y. Sugioka (1983). Restrictive if/when clauses. *Linguistics and Philosophy*, 6, 225-258.

Fauconnier, G. (1985). *Mental spaces: Aspects of Meaning Construction in Natural Language*. MIT Press, Cambridge, MA and London.

Fauconnier, G. (1997). *Mappings in Thought and Language*. Cambridge University Press, Cambridge.

Fillmore, C. (1982). Towards a descriptive framework for spatial deixis. In: *Speech, Place, and Action: Studies in Deixis and Related Topics* (R. Jarvella and W. Klein, eds.), pp. 31-59. Wiley, Chichester.

Fillmore, C. (1985). Frames and the semantics of understanding. *Quaderni di Semantica*, **6**, 222-254.

Fillmore, C. and B. Atkins (1992). Toward a frame-based lexicon: The semantics of RISK and its neighbours. In: *Frames, Fields, and Contrasts: New Essays in Semantic and Lexical Organisation* (A. Lehrer and E. Kittay, eds.), pp. 75-102. Erlbaum, Hillsdale, NJ.

Fillmore, C., P. Kay and M. O'Connor (1988). Regularity and idiomaticity in grammatical constructions: The case of *let alone*. *Language*, **64**, 501-538.

Flavell, J. (1986). The development of children's knowledge about the appearance-reality distinction. *American Psychologist*, **41**, 418-425.

Flavell, J., E. Flavell and F. Green (1987). Young children's knowledge about the apparent-real and pretend-real distinctions. *Developmental Psychology*, **23**, 816-822.

Fleischman, S. (1982). *The Future in Thought and Language*. Cambridge University Press, Cambridge.

Fodor, J. (1975). *The Language of Thought*. Crowell, New York.

Fodor, J. (1981). The present status of the innateness controversy. In: *Representations,* pp. 257-316. Harvester Press, Brighton.

Fodor, J. (1983). *The Modularity of Mind.* MIT Press, Cambridge, MA.

Fodor, J. (1990). *A Theory of Content (And Other Essays).* MIT Press, Cambridge, MA.

Fodor, J. (1998). *Concepts: Where Cognitive Science Went Wrong.* Clarendon Press, Oxford.

Fodor, J. and E. Lepore (1998). The emptiness of the lexicon: Critical reflections on J. Pustejovsky's *The generative lexicon. Linguistic Inquiry*, **29**, 269-288.

Foolen, A. (1992). Review of Sweetser (1990). *Lingua*, **88**, 76-86.

Franks, B. (1995). Sense-generation: a 'quasi-classical' approach to concepts and concept combination. *Cognitive Science*, **19**, 441-505.

Franks, B. and N. Braisby (1990). Sense generation or how to make a mental lexicon flexible. In: *Proceedings of the 12th Annual Conference of the Cognitive Science Society*. MIT Press, Cambridge, MA.

Frith, U. (1989). *Autism: Explaining the Enigma*. Blackwell, Oxford.

Gazdar, G. (1979). *Pragmatics: Implicature, Presupposition, and Logical Form.* Academic Press, New York.

Gazdar, G., G. Pullum and I. Sag (1982). Auxiliaries and related phenomena in a restrictive theory of grammar. *Language*, **58**, 591-638.

Gee, J. and I. Savasir (1985). On the use of will and gonna: Toward a description of activity-types for child language. *Discourse Processes*, **8**, 143-175.

Gerhardt, J. (1998). The meaning and use of the modals HAFTA, NEEDTA and WANNA in children's speech. *Journal of Pragmatics*, **16**, 531-590.

Gibbs, R. (1994). *The Poetics of Mind.* Cambridge University Press, Cambridge.

Gibbs, R. and J. Moise (1997). Pragmatics in understanding what is said. *Cognition*, **62**, 51-74.

Givón, T. (1978). Negation in language: Pragmatics, function, ontology. In: *Radical Pragmatics* (P. Cole, ed.), pp. 69-112. Academic Press, New York.

Goldberg, A. (1995). *Constructions: A Construction-Grammar Approach to Argument Structure.* Chicago University Press, Chicago.

Goodman, N. (1955). *Fact, Fiction, and Forecast.* Harvard University Press, Cambridge, MA.

Goodman, N. (1970). Seven strictures on similarity. In: *Experience and Theory* (L. Foster and J. Swanson, eds.). University of Massachusetts Press, Amherst.

Goossens, L. (1982). On the development of the modals and of the epistemic function in English. In: *Papers from the 5th International Conference on Historical Linguistics* (A. Alqvist, ed.), pp. 74-84. Benjamins, Amsterdam.

Gopnik, A. (1993). How we know our minds: The illusion of first-person knowledge of intentionality. *Behavioral and Brain Sciences*, **16**, 1-14.

Gopnik, A. and A. Meltzoff (1997). *Words, Thoughts, and Theories.* MIT Press, Cambridge, MA and London.

Gopnik, A. and H. Wellman (1994). The theory theory. In: *Mapping the Mind: Domain Specificity in Cognition and Culture* (L. Hirschfeld and S. Gelman, eds.), pp. 257-293. Cambridge University Press, Cambridge.

Gopnik, A. and V. Slaughter (1991). Young children's understanding of changes in their mental states. *Child Development*, **62**, 98-110.

Goshke, T. and D. Koppelberg (1992). The concept of representation and the representation of concepts in connectionist models. In: *Philosophy and Connectionist Theory* (W. Ramsey, S. Stich and D. Rumelhart, eds.), pp. 129-161. Erlbaum, Hillsdale, NJ.

Greene, J. and Wason P. (1970). Negation: A rejoinder to Wales and Grieve. *Perception and Psychophysics*, **8**, 238-239.

Grice, P. (1975). Logic and conversation. In: *Syntax and Semantics,* Vol. 3: Speech Acts (P. Cole and J. Morgan, eds.), pp. 41-58. Academic Press, New York. Reprinted in: *Pragmatics: A Reader* (S. Davis, ed., 1991), pp. 305-315. Oxford University Press, Oxford.

Grice, P. (1981). Presupposition and conversational implicature. In: *Radical Pragmatics* (P. Cole, ed.), pp. 183-198. Academic Press, New York.

Grice, P. (1989). *Studies in the Ways of Words.* Harvard University Press, Cambridge, MA.

Grober, E. (1976). *Polysemy: Its Implications for a Psychological Model of Meaning.* Doctoral dissertation, Johns Hopkins University.

Groefsema, M. (1995). *Can, may, must* and *should*: A relevance theoretic account. *Journal of Linguistics*, **31**, 53-79.

Groenendijk, J. and M. Stokhof (1975). Modality and conversational information. *Theoretical Linguistics*, **2**, 61-112.

Guo, J. (1994). *Social Interaction, Meaning, and Grammatical Form: Children's Development and Use of Modal Auxiliaries in Mandarin Chinese*. Doctoral dissertation, University of California at Berkeley.

Haegeman, L. (1983). *The Semantics of* Will *in Present-Day British English: A Unified Acccount.* Paleis der Academiën, Brussels.

Hale, K. and S. Keyser (1993). On argument structure and the lexical expressions of syntactic relations. In: *The View from Building 20: Essays in Honour of Sylvain Bromberger* (K. Hale and S. Keyser, eds.), pp. 53-110. MIT Press, Cambridge, MA.

Halliday, M. (1970). Functional diversity in language as seen from a consideration of modality and mood in English. *Foundations of Language*, **6**, 322-361.

Hare, R. (1970). Meaning and speech acts. *Philosophical Review*, **79**. Reprinted in: *Practical Inferences* (1971), pp. 74-93. Macmillan, London.

Harman, G. (1976). Practical reasoning. *Review of Metaphysics*, **29**, 431-463.

Harris, P. and M. Núñez (1997). Children's understanding of permission and obligation. In: *Piaget, Vygotsky and Beyond* (L. Smith, J. Dockrell and P. Tomlinson, eds.), pp. 211-223. Routledge, London and New York.

Heim, I. (1982). *The Semantics of Definite and Indefinite Noun Phrases.* Doctoral dissertation, University of Massachusetts, Amherst.

Heine, B. (1995). Agent-oriented vs. epistemic modality: Some observations on German modals. In: *Modality in Grammar and Discourse* (J. Bybee and S. Fleischman, eds.), pp. 17-53. Benjamins, Amsterdam.

Heine, B., U. Claudi and F. Hünnemeyer (1991). *Grammaticalization: A Conceptual Framework.* University of Chicago Press, Chicago and London.

Hirst, W. and J. Weil (1982). Acquisition of epistemic and deontic meaning of modals. *Journal of Child Language*, **9**, 659-666.

Hofmann, T. (1966). Past tense replacement and the modal system. Reprinted in: *Syntax and Semantics* (J. McCawley, ed.), Vol. 7. Academic Press, New York.

Horn, L. (1972). *On the Semantic Properties of Logical Operators in English.* Doctoral dissertation, UCLA. IULC, Bloomington.

Horn, L. (1985). Metalinguistic negation and pragmatic ambiguity. *Language*, **61**, 121-174.

Horn, L. (1989). *A Natural History of Negation.* University of Chicago Press, Chicago and London.

Huddleston, R. (1974). Further remarks on the analysis of auxiliaries as main verbs. *Foundations of Language*, **11**, 215-229.

Hughes, G. and M. Cresswell (1968). *An Introduction to Modal Logic*. Methuen, London.

Iatridou, S. (1990). The past, the possible and the evident. *Linguistic Inquiry*, **21**, 123-129.

Ifantidou, E. (1994). *Parentheticals and Relevance*. Doctoral dissertation, University of London. To appear from Benjamins.

Jackendoff, R. (1972). *Semantic Interpretation in Generative Grammar*. MIT Press, Cambridge, MA.

Jackendoff, R. (1975). On belief contexts. *Linguistic Inquiry* 6/1: 53-93.

Jackendoff, R. (1983). *Semantics and Cognition*. MIT Press, Cambridge, MA.

Jackendoff, R. (1990). *Semantic Structures*. MIT Press, Cambridge, MA.

Johnson-Laird, P. (1982). Formal semantics and the psychology of meaning. In: *Processes, Beliefs, and Questions* (S. Peters and E. Saarinen, eds.), pp. 1-68. Reidel, Dordrecht.

Joos, M. (1964). *The English Verb: Form and Meanings*. University of Wisconsin Press, Madison and Milwaukee, Wisc.

Kamp, H. (1978). Semantics versus pragmatics. In: *Formal Semantics and Pragmatics for Natural Languages* (F. Guenthner and S. Schmidt, eds.), pp. 255-287. Reidel, Dordrecht.

Karttunen, L (1972). Possible and must. In *Syntax and Semantics* (P. Kimball, ed.), Vol. 1, pp. 1-20. Seminar Press, New York.

Kay, P. (1997). *Words and the Grammar of Context*. CSLI Publications, Stanford.

Kempson, R. (1977). *Semantic Theory*. Cambridge University Press, Cambridge.

Kempson, R. (1980). Ambiguity and word-meaning. In: *Studies in English Linguistics for Randolph Quirk* (S. Greenbaum, G. Leech and J. Svartvik, eds.), pp. 7-16. Longman, London and New York.

Kempson, R. (1986). Ambiguity and the semantics-pragmatics distinction. In: *Meaning and Interpretation* (C. Travis, ed.), pp. 77-103. Blackwell, Oxford.

Klinge, A. (1993) The English modal auxiliaries: From lexical semantics to utterance interpretation. *Journal of Linguistics*, **29**, 315-357.

Kneale, W. and M. Kneale (1962). *The Development of Logic*. Oxford University Press, Oxford.

Kratzer, A. (1977). What 'must' and 'can' must and can mean. *Linguistics and Philosophy*, **1**, 337-355.

Kratzer, A. (1980). Possible-world semantics and psychological reality. *Linguistische Berichte*, **66**, 1-14.

Kratzer, A. (1981a). The notional category of modality. In: *Worlds, Words, and Contexts* (H.-J. Eikmeyer and H. Rieser, eds.), pp. 38-74. de Gruyter, Berlin.

Kratzer, A. (1981b). Partition and revision: The semantics of counterfactuals. *Journal of Philosophical Logic*, **10**, 201-216.

Kratzer, A. (1989). An investigation of the lumps of thought. *Linguistics and Philosophy*, **12**, 607-53.

Kratzer, A. (1991). Modality. In: *Semantics: An International Handbook of Contemporary Research* (A. von Stechow and D. Wunderlich, eds.), pp. 639-650. De Gruyter, Berlin.

Kuczaj, S. (1977). Old and new forms, old and new meanings: The form-function hypothesis revisited. Paper presented at the meeting of the Society for Research on Child Development, New Orleans.

Kuczaj, S. and M. Maratsos (1975). What a child CAN do before he WILL. *Merill-Palmer Quarterly*, **21**, 89-111.

Kuczaj, S. and M. Maratsos (1983). Initial verbs of yes-no questions: A different kind of general grammatical category. *Developmental Psychology*, **19**, 440-444.

Kuroda, S.-Y. (1982). Indexed predicate calculus. *Journal of Semantics*, **1**, 43-59.

Lakoff, G. (1987). *Women, Fire, and Dangerous Things*. University of Chicago Press, Chicago and London.

Lakoff, R. (1972). The pragmatics of modality. In: *Papers from the 8th Regional Meeting of CLS*, pp. 229-246. CLS, Chicago.

Langacker, R. (1987). *Foundations of Cognitive Grammar*. Vol. 1. Stanford University Press, Stanford.

Langacker, R. (1988). A usage-based model. In: *Topics in Cognitive Linguistics* (B. Rudzka-Ostyn, ed.), pp. 127-161. Benjamins, Amsterdam.

Langacker, R. (1990). Subjectification. *Cognitive Linguistics*, **1**, 5-38.

Langacker, R. (1991). *Foundations of Cognitive Grammar*. Vol. 2. Stanford University Press, Stanford.

Leech, G. (1987). *Meaning and the English Verb*. 2nd ed. Longman, London.

Leech, G. and J. Coates (1980). Semantic indeterminacy and the modals. In: *Studies in English Linguistics for Randolph Quirk* (S. Greenbaum, G. Leech and J. Svartvik, eds.), pp. 79-90. Longman, London.

Leekham, S. and J. Perner (1991). Does the autistic child have a metarepresentational deficit? *Cognition*, **40**, 203-218.

Lehrer, A. and E. Kittay, eds. (1992). *Frames, Fields, and Contrasts: New Essays in Semantic and Lexical Organisation*. Erlbaum, Hillsdale, NJ.

Leslie, A. (1988). Some implications of pretense for mechanisms underlying the child's theory of mind. In: *Developing Theories of Mind* (J. Astington, P. Harris and D. Olson, eds.), pp. 19-46. Cambridge University Press, New York.

Leslie, A. (1991). The Theory of Mind impairment in autism: Evidence for a modular mechanism for development? In: *Natural Theories of Mind* (A. Whiten, ed.), pp. 63-78. Blackwell, Oxford.

Leslie, A. (1994). ToMM, ToBy, and agency: Core architecture and domain specificity. In: *Mapping the Mind: Domain Specificity in Cognition and Culture* (L. Hirschfeld and S. Gelman, eds.), pp. 119-148. Cambridge University Press, Cambridge.

Leslie, A. (1995). Pretending and believing: Issues in the theory of ToMM. In: *Cognition* on Cognition (J.. Mehler and S. Franck, eds.), pp. 193-220. MIT Press, Cambridge, MA.

Leslie, A. (forth.). How to acquire a 'representational theory of mind'. In: *Metarepresentation: Vancouver Studies in Cognitive Science*, Vol. 10 (D. Sperber and S. Davis, eds.). Oxford University Press, Oxford.

Leslie, A. and D. Roth (1993). What autism teaches us about metarepresentation. In: *Understanding Other Minds: Perspectives from Autism* (S. Baron-Cohen, H. Tager-Flusberg and D. Cohen, eds.), pp. 83-111. Oxford University Press, Oxford.

Leslie, A. and F. Happé (1989). Autism and ostensive communication: The relevance of metarepresentation. *Development and Psychopathology*, **1**, 205-212.

Leslie, A. and L. Thaiss (1992). Domain specificity in conceptual development: Evidence from autism. *Cognition*, **43**, 225-251.

Levin, B. (1993). *English Verb Classes and Alternations: A Preliminary Investigation.* University of Chicago Press, Chicago.

Levin, B. and S. Pinker, eds. (1991). *Lexical and Conceptual Semantics.* MIT Press, Cambridge, MA.

Levinson, S. (1983). *Pragmatics.* Cambridge University Press, Cambridge.

Levinson, S. (1999). *Presumptive Meanings: The Theory of Generalized Conversational Implicature.* MIT Press, Cambridge, MA.

Lewis, D. (1973a). *Counterfactuals.* Blackwell, Oxford.

Lewis D. (1973b). Counterfactual dependence and time's arrow. Repr. in: *Philosophical Papers*, Vol. II (1986), pp. 32-66. New York and Oxford University Press, Oxford.

Lewis, D. (1975). Adverbs of quantification. In: *Formal Semantics of Natural Language* (E. Keenan, ed.), pp. 3-15. London and Cambridge University Press, New York.

Lewis, D. (1979). Scorekeeping in a language game. *Journal of Philosophical Language*, **8**, 339-359. Reprinted in: *Pragmatics: A Reader* (S. Davis, ed., 1991), pp. 416-427. Oxford University Press, New York.

Lewis, D. (1981). Counterfactuals and comparative possibility. In: *Ifs: Conditionals, Belief, Decision, Chance, and Time* (W. Harper, R. Stalnaker and G. Pearce, eds.), pp. 57-85. Reidel, Dordrecht.

Lewis, D. (1986). *On the Plurality of Worlds.* Blackwell, Oxford.

Lieber, R. (1980). *On the Organisation of the Lexicon.* Doctoral dissertation, MIT. IULC Bloomington.

Lukes, S. and I. Galnoor (1987). *No Laughing Matter: A Collection of Political Jokes.* Penguin, London.

Lyons, J. (1977). *Semantics.* 2 vols. Cambridge University Press, Cambridge.

Matsui, T. (1995). *Relevance and Bridging.* Doctoral dissertation, University of London. To appear from Benjamins.

Mayer, R. (1990). Abstraction, context, and perspectivization - evidentials in Discourse Semantics. *Theoretical Linguistics*, **16**, 101-163.

McCawley, J. (1981). *Everything that Linguists Have Always Wanted to Know about Logic (But Were Ashamed to Ask).* 2nd ed. University of Chicago Press, Chicago and London.

Moore, C. and J. Davidge (1989). The development of mental terms: Pragmatics or semantics? *Journal of Child Language*, **16**, 633-641.

Moore, C., D. Bryant and D. Furrow (1989). Mental terms and the development of certainty. *Child Development*, **60**, 167-171.

Moore, C., K. Pure and D. Furrow (1990). Children's understanding of the modal expression of speaker certainty and uncertainty and its relation to the development of a representational theory of mind. *Child Development*, **61**, 722-730.

Moshman, D. (1990). The development of metalogical understanding. In: *Reasoning, Necessity, and Logic: Developmental Perspectives* (W. Overton, ed.), pp. 205-226. Erlbaum, Hillsdale, NJ.

Neale, S. (1990). *Descriptions.* MIT Press, Cambridge, MA and London.

Newmeyer, F. (1970). The root modal: Can it be transitive? In: *Festschrift for Robert B. Lees.* Linguistic Research Inc.

Noh, E. (1998). The Semantics and Pragmatics of Metarepresentations in English: A Relevance-Theoretic Approach. Doctoral dissertation, University of London. To appear from Benjamins.

Noveck, I., S. Ho and M. Sera (1996). Children's understanding of epistemic modals. *Journal of Child Language*, **23**, 621-643.

Nunberg, G. (1978). *The Pragmatics of Reference.* IULC, Bloomington.

Nunberg, G. (1979). The non-uniqueness of semantic solutions: Polysemy. *Linguistics and Philosophy*, **3**, 143-184.

Nunberg, G. and A. Zaenen (1992). Systematic polysemy in lexicology and lexicography. In: *Proceedings of EURALEX* (H. Tommola, K. Varantola, T. Tolonen and J. Schopp, eds.), Vol. 2, pp. 387-396. University of Tampere, Tampere, Finland.

Nuyts, J. (1993). Epistemic modal adverbs and adjectives and the layered representation of conceptual and linguistic structure. *Linguistics*, **31**, 933-969.

O'Neill, D. and A. Gopnik (1991). Young children's ability to identify the sources of their beliefs. *Developmental Psychology*, **27**, 390-397.

Osherson, D. and E. Markman (1975). Language and the ability to evaluate contradictions and tautologies. *Cognition*, **3**, 213-222.

Ostler, N. and B. Atkins (1992). Predictable meaning shift: some linguistic properties of lexical implication rules. In: *Lexical Semantics and Knowledge Representation. Proceedings of the First SIGLEX Workshop: Lexical Semantics and Knowledge Representation* (J. Pustejovsky and S. Bergler, eds.), pp. 87-100. Springer, Berlin.

Overton, W. (1990). Competence and procedures: Constraints on the development of logical reasoning. In: *Reasoning, Necessity, and Logic: Developmental Perspectives* (W. Overton, ed.), pp. 1-32. Erlbaum, Hillsdale, NJ.

Palmer, F. (1986). *Mood and Modality*. Cambridge University Press, Cambridge.

Palmer, F. (1990). *Modality and the English Modals*. 2nd ed. Longman, London and New York.

Palmer, F. (1995). Negation and the modals of possibility and necessity. In: *Modality in Grammar and Discourse* (J. Bybee and S. Fleischman, eds.), pp. 453-471. Benjamins, Amsterdam.

Panman, O. (1982). Homonymy and polysemy. *Lingua*, **58**, 105-136.

Papafragou, A. (1995). Metonymy and relevance. *UCL Working Papers in Linguistics*, **7**, 141-175. Revised version as Papafragou (1996b).

Papafragou, A. (1996a). On generics. *UCL Working Papers in Linguistics*, **8**, 165-198.

Papafragou, A. (1996b). On metonymy. *Lingua*, **99**, 169-195.

Papafragou, A. (1998a). *Modality and the Semantics-Pragmatics Interface*. Doctoral Dissertation, University of London.

Papafragou, A. (1998b). Kinds of modality <in Greek>. In: *Studies in Greek Linguistics (Proceedings from the 18th Annual Meeting of the Department of Linguistics, University of Thessaloniki)*, pp. 414-427. Kyriakidis Publishers, Thessaloniki.

Papafragou, A. (1998c). Modality and metarepresentation. In: *Proceedings from the 22nd Annual Boston University Conference on Language Development*. Cascadilla Press, Somerville, MA.

Papafragou, A. (1998d). Modality and semantic underdeterminacy. In: *Current Issues in Relevance Theory* (V. Rouchota and A. Jucker, eds.), pp. 239-274. Benjamins, Amsterdam.

Papafragou, A. (1998e). The acquisition of modality: Implications for theories of semantic representation. *Mind and Language*, **13**, 370-399.

Papafragou, A. (1998f). Inference and word meaning: The case of modal auxiliaries. *Lingua*, **105**, 1-47.

Papafragou, A. (2000). Indirect requests: A rethink. Under revision.

Partee, B. (1977). Possible worlds semantics and linguistic theory. *Monist*, **60**, 303-326.

Partee, B. (1989). Binding implicit variables in quantified contexts. In: *Papers from the 25th Regional Meeting of CLS*, pp. 342-365. CLS, Chicago.

Partee, B. (1995a). Quantificational structure and compositionality. In: *Quantification in Natural Languages* (E. Bach, E. Jelinek, A. Kratzer and B. Partee, eds.), Vol. II, pp. 541-601. Kluwer, Dordrecht.

Partee, B. (1995b). Lexical semantics and compositionality. In: *An invitation to Cognitive Science* (L. Gleitman and M. Liberman, eds.), Vol. 1: *Language*, pp. 311-360. MIT Press, Cambridge, MA.

Pea, R., R. Mawby and S. MacCain (1982). World-making and world-revealing: Semantics and pragmatics of modal auxiliary verbs during the third year of life. Paper presented at the 7th Annual Boston Conference on Child Language Development, October.

Pelletier, F. and L. Schubert (1989). Mass expressions. In: *Handbook of Philosophical Logic* (D. Gabbay and F. Guenthner, eds.), Vol. 4, pp. 327-407. Reidel, Dordrecht.

Perkins, M. (1983). *Modal Expressions in English*. Frances Pinter, London.

Perlmutter, D. (1971). *Deep and Surface Structure Constraints in Syntax*. Holt, Rinehart and Winston, New York.

Perner, J. (1991). *Understanding the Representational Mind*. MIT Press, Cambridge, MA.

Perner, J. and J. Astington (1992). The child's understanding of mental representation. In: *Piaget's Theory: Prospects and Possibilities* (H. Beilin and P. Pufall, eds.), pp. 141-160. Erlbaum, Hillsdale, NJ.

Perner, J., S. Leekam and H. Wimmer (1987). Three-year-olds' difficulty with false belief: The case for a conceptual deficit. *British Journal of Developmental Psychology*, **5**, 125-137.

Piaget, J. (1928). *Judgement and Reasoning in the Child*. Routledge, London.

Piaget, J. (1987). *Possibility and Necessity*. 2 vols. University of Minnesota Press, Minneapolis.

Piatelli-Palmarini, M., ed. (1980). *Language and Learning: The Debate between Jean Piaget and Noam Chomsky*. Routledge, London.

Picallo, C. (1990). Modal verbs in Catalan. *Natural Language and Linguistic Theory*, **8**, 285-312.

Pieraut-LeBonniec, G. (1980). *The Development of Modal Reasoning: Genesis of Necessity and Possibility Notions*. Academic Press, New York.

Pinkal, M. (1995). *Logic and Lexicon*. Reidel, Dordrecht.

Pratt, C. and P. Bryant (1990). Young children understand that looking leads to knowing (so long as they are looking into a single barrel). *Child Development*, **61**, 973-982.

Premack, D. and A. Premack (1994). Moral belief: Form versus content. In: *Mapping the Mind: Domain Specificity in Cognition and Culture* (L. Hirschfeld and S. Gelman, eds.), pp. 149-168. Cambridge University Press, Cambridge.

Pustejovsky, J. (1993). Type coercion and lexical selection. In: *Semantics and the Lexicon* (J. Pustejovsky, ed.), pp. 73-96. Kluwer, Dordrecht.

Pustejovsky, J. (1995). *The Generative Lexicon*. MIT Press, Cambridge, MA and London.

Pustejovsky, J. and B. Boguraev, eds. (1996). *Lexical Semantics: The Problem of Polysemy*. Clarendon Press, Oxford.

Pustejovsky, J. and P. Bouillon (1996). Aspectual coercion and logical polysemy. In: *Lexical Semantics: The Problem of Polysemy* (J. Pustejovsky and B. Boguraev, eds.), pp. 133-162. Clarendon Press, Oxford.

Pustejovsky, J., ed. (1993). *Semantics and the Lexicon*. Kluwer, Dordrecht.

Quine, W. V. (1940). *Mathematical Logic*. Harvard University Press, Cambridge, MA. Revised ed. 1951.

Quirk, R., S. Greenbaum, G. Leech and J. Svartvik (1985). *A Comprehensive Grammar of the English Language*. Longman, London and New York.

Recanati, F. (1987). *Meaning and Force*. Cambridge University Press, Cambridge.

Recanati, F. (1989). The pragmatics of what is said. *Mind and Language*, **4**, 295-329.

Recanati, F. (1993). *Direct Reference: From Language to Thought*. Blackwell, Oxford.

Recanati, F. (1996). Domains of discourse. *Linguistics and Philosophy*, **19**, 445-475.

Rice, S. (1992). Polysemy and lexical representation: The case of three English prepositions. In: *Proceedings of the 14th Annual Conference of the Cognitive Science Society*, pp. 89-94. Erlbaum, Hillsdale, NJ.

Roberts, C. (1989). Modal subordination and pronominal anaphora in discourse. *Linguistics and Philosophy*, **12**, 683-721.

Roberts, C. (1995). Domain restriction in dynamic semantics. In: *Quantification in Natural Languages* (E. Bach, E. Jelinek, A. Kratzer and B. Partee. eds.), Vol. II, pp. 661-700. Kluwer, Dordrecht.

Roberts, I. (1985). Agreement parameters and the development of English modal auxiliaries. *Natural Language and Linguistic Theory*, **3**, 21-58.

Ross, J. (1969). Auxiliaries as main verbs. In: *Studies in Philosophical Linguistics* (W. Todd, ed.), Series I. Great Expectations Press, Evanston, IL.

Rouchota, V. (1994a). *The Semantics and Pragmatics of the Subjunctive in Modern Greek: A Relevance-Theoretic Approach*. Doctoral dissertation, University of London. To appear from Benjamins.

Rouchota, V. (1994b). On indefinite descriptions. *Journal of Linguistics*, **30**, 441-475.

Rouchota, V. and A. Jucker, eds. (1998). *Current Issues in Relevance Theory*. Benjamins, Amsterdam.

Ruhl, C. (1989). *On Monosemy: A Study in Linguistic Semantics*. State University of New York Press, Stony Brook.

Russell, J. (1982). The child's appreciation of the necessary truth and the necessary falseness of propositions. *British Journal of Psychology*, **73**, 253-266.

Ruwet, N. (1991). Raising and control revisited. In: *Syntax and Human Experience* (J. Goldsmith, ed. and transl.), pp. 56-81. University of Chicago Press, Chicago and London.

Saeed, J. (1997). *Semantics*. Blackwell, Oxford.

Sag, I. and A. Szabolski, eds. (1992). *Lexical Maters*. CSLI Lecture Notes. Stanford, CA, CSLI.

Sanders, J. and W. Spooren (1997). Perspective, subjectivity, and modality from a cognitive linguistic point of view. In: *Discourse and Perspective in Cognitive Linguistics* (W. Liebert, G. Redeker and L. Waugh, eds.), pp. 85-112. Benjamins, Amsterdam.

Schubert, L. and F. Pelletier (1989). Generically speaking. In: *Properties, Types and Meaning* (G. Chierchia, B. Partee and R. Turner, eds.), Vol. 2, pp. 193-268. Kluwer, Dordrecht.

Searle, J. (1969). *Speech Acts*. Cambridge University Press, Cambridge.

Searle, J. (1983). *Intentionality*. Cambridge University Press, Cambridge.

Sellars, W. (1954). Presupposing. *Philosophical Review*, **63**, 197-215.

Shatz, M. and S. Wilcox (1991). Constraints on the acquisition of English modals. In: *Perspectives on Language and Thought* (S. Gelman and J. Byrnes, eds.), pp. 319-353. Cambridge University Press, Cambridge.

Shatz, M., D. Billman and I. Yaniv (1986). Early occurrences of English auxiliaries in children's speech. University of Michigan ms., Ann Arbor.

Shatz, M., H. Grimm, S. Wilcox and K. Niemeier-Wind (1989). The uses of modal expressions in conversations between German and American mothers and their two-year-olds. Paper presented at the biennial meeting of the Society for Research in Child Development, Kansas City, MO.

Shatz, M., H. Grimm, S. Wilcox and K. Niemeier-Wind (1990). Modal expressions in German and American mother-child conversations: Implications for input theories of language acquisition. University of Michigan ms. Ann Arbor.

Shatz, M., H. Wellman and S. Silber (1983). The acquisition of mental verbs: A systematic investigation of the first reference to mental state. *Cognition*, **14**, 301-321.

Shepherd, S. (1982). From deontic to epistemic: An analysis of modals in the history of English, creoles, and language acquisition. In: *Papers from the 5th International Conference on Historical Linguistics* (A. Alqvist, ed.), pp. 316-323. Benjamins, Amsterdam.

Shepherd, S. (1993). The acquisition of modality in Antiguan Creole. In: *Modality in Language Acquisition* (N. Dittmar and A. Reich, eds.), pp. 171-185. de Gruyter, Berlin.

Slobin, D. and A. Aksu (1982). Tense, aspect, and modality in the use of the Turkish evidential. In: *Tense and Aspect: Between Semantics and Pragmatics* (P. Hopper, ed.), pp. 185-200. Benjamins, Amsterdam.

Smith, L. (1993). *Necessary Knowledge: Piagetian Perspectives on Constructivism.* Erlbaum, Hillsdale, NJ.

Smith, N. (1983). On interpreting conditionals. *Australasian Journal of Linguistics*, **3**, 1-24.

Smith, N. (1989). *The Twitter Machine.* Blackwell, Oxford.

Smith, N. and A. Smith (1988). A relevance-theoretic account of conditionals. In; *Language, Speech and Mind: Essays in Honour of Victoria Fromkin* (L. Hyman and C. Li, eds.), PP. 322-352. Routledge, London.

Smoczynska, M. (1993). The acquisition of Polish modal verbs. In: *Modality in Language Acquisition* (N. Dittmar and A. Reich, eds.), pp. 145-169. de Gruyter, Berlin.

Sophian, C. and S. Somerville (1988). Early developments in logical reasoning: considering alternative possibilities. *Cognitive Development*, **3**, 183-222.

Sperber, D. (1994a). Understanding verbal understanding. In: *What Is Intelligence?* (J. Khalfa, ed.), pp. 179-198. Cambridge University Press, Cambridge.

Sperber, D. (1994b). The modularity of thought and the epidemiology of representations. In: *Mapping the Mind: Domain Specificity in Cognition and Culture* (L. Hirschfeld and S. Gelman, eds.), pp. 39-67. Cambridge University Press, Cambridge.

Sperber, D. (1996). *Explaining Culture: A Naturalistic Approach.* Blackwell, Oxford.

Sperber, D. (1997). Intuitive and reflective beliefs. *Mind and Language*, **12**, 67-83.

Sperber, D. and D. Wilson (1984). Draft of *Relevance,* Chapter VII: Speech Acts and Propositional Attitudes. UCL ms.

Sperber, D. and D. Wilson (1985/6). Loose talk. *Proceedings of the Aristotelian Society*, **86**, 153-171. Reprinted in: *Pragmatics: A Reader* (S. Davis, ed., 1991), pp. 540-549. Oxford University Press, Oxford.

Sperber, D. and D. Wilson (1986). *Relevance: Communication and Cognition.* Blackwell, Oxford. 2nd. ed. 1995.

Sperber, D. and D. Wilson (1987). Précis of Relevance. *Behavioural and Brain Sciences*, **10**, 697-710.

Sperber, D. and D. Wilson (1989). On verbal irony. *UCL Working Papers in Linguistics*, **1**, 96-117.

Sperber, D. and D. Wilson (1990). Rhetoric and relevance. In: *The Ends of Rhetoric: History, Theory, Practice* (D. Wellbery and J. Bender, eds.), pp. 140-155. Stanford University Press, Stanford, CA.

Sperber, D. and D. Wilson (1997). The mapping between the mental and the public lexicon. *UCL Working Papers in Linguistics*, **9**, 107-125. Reprinted in: *Language and Thought: Interdisciplinary Themes* (P. Carruthers and J. Boucher, eds., 1998), pp. 184-200. Cambridge University Press, Cambridge.

Stalnaker, R. (1968). A theory of conditionals. In: *Studies in Logical Theory* (N. Rescher, ed.). Reprinted in: *Ifs: Conditionals, Belief, Decision, Chance, and Theory* (W. Harper, R. Stalnaker and G. Pearce, eds., 1981), pp. 41-55. Reidel, Dordrecht.

Stalnaker, R. (1970). Probability and conditionals. Reprinted in *Ifs: Conditionals, Belief, Decision, Chance, and Theory* (W. Harper, R. Stalnaker and G. Pearce, eds., 1981), pp. 107-128. Reidel, Dordrecht.

Stalnaker, R. (1980). A defense of conditional excluded middle. Reprinted in: *Ifs: Conditionals, Belief, Decision, Chance, and Theory* (W. Harper, R. Stalnaker and G. Pearce, eds., 1981), pp. 87-104. Reidel, Dordrecht.

Stalnaker, R. (1986). Possible worlds and situations. *Journal of Philosophical Logic*, **15**, 109-23.

Steedman, M. (1977). Verbs, time, and modality. *Cognitive Science*, **1**, 216-234.

Stephany, U. (1979). Modality. In: *Language Acquisition* (P. Fletcher and M. Garman, eds.), pp. 375-400. Cambridge University Press, Cambridge. 2nd ed. 1986.

Stephany, U. (1993). Modality in first-language acquisition: The state of the art. In: *Modality in Language Acquisition* (N. Dittmar and A. Reich, eds.), pp. 133-144. de Gruyter, Berlin and New York.

Sweetser, E. (1986). Polysemy vs. abstraction: Mutually exclusive or complementary? In: *Proceedings of the 12th Annual Meeting of the BLS* (K. Nikiforidou, M. VanClay, M. Niepokiy and D. Feder, eds.), pp. 527-538. BLS, Berkeley, CA.

Sweetser, E. (1988). Grammaticalization and semantic bleaching. In: *Proceedings of the 14th Annual Meeting of the BLS* (S. Axmaker, A. Jaisser and H. Singmaster, eds.), pp. 389-405. BLS, Berkeley, CA.

Sweetser, E. (1990). *From Etymology to Pragmatics*. Cambridge University Press, Cambridge.

Tager-Flusberg, H. (1993). What language reveals about the understanding of minds in children with autism. In: *Understanding Other Minds: Perspectives from Autism* (S.

Baron-Cohen, H. Tager-Flusberg and D. Cohen, eds.), pp. 138-157. Oxford University Press, Oxford.

Tager-Flusberg, H. (1997). Language acquisition and theory of mind: Contributions from the study of autism. In: *Research on Communication and Language Development* (L. Adamson and M. Romski, eds.), pp. 133-158. Paul Brookes Publishing, Baltimore, MD.

Talmy, L. (1988). Force dynamics in language and cognition. *Cognitive Science*, **2**, 49-100.

Tanaka, T. (1990). Semantic changes of *can* and *may*: differentiation and implication. *Journal of Linguistics*, **266**, 89-123.

Taylor, J. (1995). *Linguistic Categorisation.* Oxford University Press, Oxford. 2nd ed. (1st ed. 1989).

Taylor, M. (1988). The development of children's understanding of the seeing-knowing distinction. In: *Developing Theories of Mind* J. Astington, P. Harris and D. Olson, eds.), pp. 207-225. Cambridge University Press, New York.

Tomasello, M. and A. Kruger (1992). Acquiring verbs in ostensive and non-ostensive contexts. *Journal of Child Language*, **19**, 311-333.

Traugott, E. (1982). From propositional to textual and expressive meanings: Some semantic-pragmatic aspects of grammaticalization. In: *Perspectives on Historical Linguistics* (W. Lehman and Y. Malkiel, eds.), pp. 245-271. Benjamins, Amsterdam.

Traugott, E. (1986). From polysemy to internal semantic reconstruction. In *Proceedings of the 12th Annual Meeting of the BLS* (K. Nikiforidou, M. van Clay, M. Niepokiy and D. Feder, eds.), pp. 539-550. BLS, Berkeley, CA.

Traugott, E. (1988). Pragmatic strengthening and grammaticalization. In: *Proceedings of the 14th Annual Meeting of the BLS* (S. Axmaker, A. Jaisser and H. Singmaster, eds.), pp. 406-416. BLS, Berkeley, CA.

Traugott, E. (1989). On the rise of epistemic meanings in English: An example of subjectification in semantic change. *Language*, **65**, 31-55.

Traugott, E. (1995). Subjectification in grammaticalisation. In: *Subjectivity and Subjectivisation: Linguistic Perspectives* (D. Stein and S. Wright, eds.), pp. 31-54. Cambridge University Press, Cambridge.

Traugott, E. and B. Heine, eds. (1991). *Approaches to Grammaticalization.* 2 vols. Benjamins, Amsterdam.

Traugott, E. and E. König (1991). The semantics-pragmatics of grammaticalization revisited. In *Approaches to Grammaticalization* (E. Traugott and B. Heine, eds.), Vol. 1, pp. 189-218. Benjamins, Amsterdam.

Travis, C. (1981). *The True and the False: The Domain of the Pragmatic*. Benjamins, Amsterdam.

Travis, C. (1985). On what is strictly speaking true. *Canadian Journal of Philosophy*, **15**, 187-229.

Tregidgo, P. (1982). MUST and MAY: Demand and permission. *Lingua*, **56**, 75-92.

Tsohatzidis, S., ed. (1990). *Meanings and Prototypes: Studies in Linguistic Categorization*. Routledge, London and New York.

Turner, K., ed. (1999). *The Semantics/Pragmatics Interface from Different Points of View*. Elsevier Science, Oxford.

Urmson, J. (1963). Parenthetical verbs. In: *Philosophy and Ordinary Language* (C. Caton, ed.), pp. 220-240. University of Illinois Press, Indiana.

van der Auwera, J. (1986). The possibilities of *may* and *can*. In: *Linguistics Across Historical and Geographical Boundaries: In Honour of Jasek Fisiak on the Occasion of his 50th Birthday* (D. Kastovsky and A. Szwedek, eds.), Vol. II, pp. 1067-1076. de Gruyter, Berlin.

Vendler, Z. (1967). Singular terms. In: *Linguistics in Philosophy*, pp. 33-69. Cornell University Press, Ithaca, NY.

von Wright, G. (1951). *An Essay in Modal Logic*. North Holland, Amsterdam.

von Wright, G. (1963). Practical inference. *Philosophical Review*, **72**, 159-179.

Walton, A. (1988). *The Pragmatics of the English Modal Verbs*. Doctoral dissertation, University of London.

Warner, A. (1993). *English Auxiliaries: Structure and History*. Cambridge University Press, Cambridge.

Wason, P. (1965). The contexts of plausible denial. *Journal of Verbal Learning and Verbal Behaviour*, **4**, 7-11.

Weinreich, U. (1964). *Webster's Third*: A critique of its semantics. *International Journal of American Linguistics*, **30**, 405-409.

Wellman, H. (1990). *The Child's Theory of Mind*. MIT Press, Cambridge, MA and London.

Wellman, H. (1993). Early understanding of mind: The normal case. In: *Understanding Other Minds: Perspectives from Autism* (S. Baron-Cohen, H. Tager-Flusberg and D. Cohen, eds.), pp. 10-39. Oxford University Press, Oxford.

Wellman, H. and D. Estes (1986). Early understanding of mental entities: A reexamination of childhood realism. *Child Development*, **57**, 910-923.

Wellman, H. and K. Bartsch (1988). Young children's reasoning about beliefs. *Cognition*, **30**, 239-277.

Wells, G. (1979). Learning and using the auxiliary verb in English. In: *Cognitive Development: Language and Thinking from Birth to Adolescence* (V. Lee, ed.), pp. 250-270. Croom Helm, London.

Wells, G. (1985). *Language Development in the Pre-School Years*. Cambridge University Press, Cambridge.

Wertheimer, R. (1972). *The Significance of Sense: Meaning, Modality and Morality*. Cornell University Press, Ithaca.

Westerstähl, D. (1985). Determiners and context sets. In: *Generalised Quantifiers in Natural Language* (J. van Benthem and A. ter Meulen, eds.), pp. 45-71. Foris, Dordrecht.

Willett, T. (1988). A cross-linguistic survey of the grammaticization of evidentiality. *Studies in Language*, **12**, 51-97.

Williams, E. (1994). A reinterpretation of the evidence for verb movement in French. In: *Verb Movement* (D. Lightfoot and N. Hornstein, eds.), pp. 189-205. Cambridge University Press, Cambridge.

Williams, J. (1992). Processing polysemous words in context: Evidence for interrelated meanings. *Journal of Psycholinguistic Research*, **21**, 193-218.

Wilson, D. (1992). Reference and relevance. *UCL Working Papers in Linguistics*, **4**, 167-191.

Wilson, D. (1993). Semantic theory. UCL Lecture notes.

Wilson, D. (1998). Linguistic structure and inferential communication. In: *Proceedings of the 16th International Congress of Linguists* (Paris, July 1997). Elsevier, Oxford.

Wilson, D. and D. Sperber (1988a). Mood and the analysis of non-declarative sentences. In: *Human Agency: Language, Duty and Value* (J. Dancy, J. Moravcsik and C. Taylor, eds.), pp. 229-324. Stanford University Press, Stanford, CA.

Wilson, D. and D. Sperber (1988b). Representation and relevance. In: *Mental Representations: The Interface between Language and Reality* (R. Kempson, ed.), pp. 133-153. Cambridge University Press, Cambridge.

Wilson, D. and D. Sperber (1993a). Linguistic form and relevance. *Lingua*, **90**, 1-25.

Wilson, D. and D. Sperber (1993b). Pragmatics and time. *UCL Working Papers in Linguistics*, **5**, 277-298. Reprinted in: *Relevance Theory: Applications and Implications* (R. Carston and S. Uchida, eds., 1997), pp. 1-22. Benjamins, Amsterdam.

Wimmer, H. and J. Perner (1983). Beliefs about beliefs: Representation and constraining function of wrong beliefs in young children's understanding of deception. *Cognition*, **13**, 103-128.

Wing, L., ed. (1976). *Early Childhood Autism*. Pergamon Press, Oxford.

Zhang, Q. (1998). Fuzziness, vagueness, generality, ambiguity. *Journal of Pragmatics*, **29**, 13-31.

Zubizarreta, M. L. (1982). *On the Relationship of the Lexicon to Syntax*. Doctoral dissertation, MIT.

Zwicky, A. and J. Sadock (1975). Ambiguity tests and how to fail them. In: *Syntax and Semantics* (J. Kimball, ed.), Vol. 4, pp. 1-36. Academic Press, New York.

Index